PROTOPLASMATOLOGIA

HANDBUCH DER PROTOPLASMAFORSCHUNG

HERAUSGEGEBEN VON

L. V. HEILBRUNN UND F. WEBER
PHILADELPHIA GRAZ

BAND II

CYTOPLASMA

B

CHEMIE

2

CYTOCHEMIE UND HISTOCHEMIE

C

pH OF PLANT CELLS
pH OF ANIMAL CELLS

WIEN
SPRINGER-VERLAG
1955

THE pH OF PLANT CELLS

BY

JAMES SMALL
BELFAST

WITH 3 FIGURES

THE pH OF ANIMAL CELLS

BY

FLOYD J. WIERCINSKI
PHILADELPHIA, PA.

WITH 7 FIGURES

WIEN
SPRINGER-VERLAG
1955

ISBN-13:978-3-211-80386-8 e-ISBN-13:978-3-7091-5451-9
DOI: 10.1007/978-3-7091-5451-9

Protoplasmatologia
II. Cytoplasma
B. Chemie
2. Cytochemie und Histochemie
c) The pH of Plant Cells

The pH of Plant Cells

By

James Small

Department of Botany, The Queen's University of Belfast

With 3 Figures

Contents

I. Introduction

Early estimations of pH values in plant cells were based upon determinations of the electromotive force (EMF) developed between a standard hydrogen electrode and the fluid expressed from crushed tissues or liberated by cutting large algal cells such as those of *Valonia, Chara* or *Nitella.* These EMF values were then translated into terms of hydrogen ion concentration on the assumption that the numerical value of the EMF in volts depended entirely upon the concentration of hydrogen ions. Then hydrogen ion concentration measured in this way was found to act in biological reactions, mainly enzyme activity, in such a way that the apparent hydrogen ion concentration was related logarithmically to the process investigated. So SÖRENSEN in 1909 introduced the pH scale based upon what he called the "hydrogen-ion exponent." He defined p_{H^+} as $\frac{1}{(\log p_{H^+})}$ (1909, p. 28, not p. 4) and p_{H^+} became pH as a matter of typographical convenience.

As far as living, uncrushed, plant cells were concerned, the available fluid was usually so minute in amount, except in special cases, that hydrogen electrode methods were not applicable to a general survey. Attention was therefore directed to pH indicators. These indicators vary in colour, usually from one colour to another colour with dichroic intermediate tints within a range of 1.6 pH units, in various parts of the pH scale which has pH 7 as neutrality with pH 6 to pH 2 as readily measurable degrees of acidity and pH 8 to pH 12 as corresponding degrees of alkalinity.

Some of the early indicators, such as neutral red, were sensitive to oxidation-reduction (H–O) or (rH) phenomena and so gave colours which were inconsistent with the colours given by indicators sensitive only to (H–OH) potentials or pH. But the range of suitable pH indicators increased and much valuable work was done by allowing plant cells to absorb relatively non-toxic indicators. Disagreements and disputes arose over the interpretation of the various tints observed, but many of these early records are still quite useful when interpreted according to SMALL's Range Indicator Method given below.

The methods of calculation of pH, from EMF in terms of volts where a one millivolt difference was considered significant, led to pH values being recorded to the third decimal place when the hydrogen electrode technique was used, but with indicators the accuracy of the estimation of pH values has seldom been taken even to the second decimal. Most indicator methods, with standardised quantities for fluid and indicator and depth of coloured fluids used for matching tints, have a gap of 0.2 pH between successive colour-standard tints, so that an intermediate reading is to the nearest first decimal. This degree of accuracy involves standardisation of indicator concentration as the same in both the fluids being compared, and also an equality of depth in these two fluids. This can be attained for living plant cells only by very special manipulation of particular materials.

The modern outlook on pH is very different from the rigid formalism and apparent precision which resulted from theoretical considerations concerning hydrogen ion concentration. The modern official standard for pH estimations is a practical, more or less arbitrary one, namely that a one-twentieth molar solution of pure potassium hydrogen phthalate at 15° C. has a pH 4.000. All other pH values, in practice, are to be referred to this as a numerical standard. There are equations for calculating the effect of temperature, and some subsidiary standards for convenience, with a recognition that the varied activity of hydrogen ions and other associated ions may lead to a precision of only ± 0.02 in the finally estimated pH value. These matters are considered later but they are mentioned here in order to show that modern standardising authorities are satisfied with a precision which begins to approximate to the first decimal with an allowable swing of .04 in the second decimal place.

Estimations of pH in plant cells may now be considered reasonably accurate if precision to the first decimal place can be attained; thus pH 5.9 ± .05 or pH 4.0 ± .05 are no longer to be regarded as mere approximations. These values are almost as nearly accurate as a chemist's results from analysis of simple solutions.

This practical outlook is the result of pH values being regarded as more or less intrinsic characteristics of various commercial or industrial materials, symptomatic of desirable or undesirable characteristics which may be connected with pH in a great variety of different ways, logically or quite empirically.

On the theoretical side, for those who prefer the niceties of mathematical solutions to pH problems, there is now the delightful book by RICCI (1952), with seven pages of Symbols and Abbreviations, beside which the Tables I, II, III given below are merely the ABC of practical very rough estimations of the possible approximate pH values for plant cells and their contents.

Table I. *R. I. M. Indicators and Colour Changes.*

Indicator	Abbrev.	Alk. Colour Range	pH	Acid Colour Range	pH
Phenol red	PR	pink to red	> 7.0	yellow	< 6.8
Bromo-thymol blue	BTB	green to blue	> 6.4	yellow	< 6.2
Bromo-cresol purple	BCP	pale blue to deep purple	> 6.2	yellow*	< 5.9
Ethyl red	ER	yellow	> 5.9	pale pink to deep red	< 5.6
Methyl red	MR	yellow	> 5.6	pale pink to deep red	< 5.2
Benzene-azo-α naphthylamine	BAN	yellow	> 4.8	pale pink to deep red	< 4.4
Bromo-cresol green	BCG	pale green to deep blue	> 4.4	yellow	< 4.0
Bromo-phenol blue	BPB	pale green to deep blue	> 4.0	yellow	< 3.4

* The yellow with BCP is stronger than the indefinite purple about 6.0–5.8.

Table II. *Ranges for pH indicated by R. I. M.*

Colours and indicators	pH Range	Notation
blue to purple BCP	> 6.2	A = > 62
yellow BCP, yellow ER	$< 5.9 > 5.9$	a = 59–59
yellow BCP, indefinite (indet) ER, yellow MR	$< 5.9 > 5.6$	b = 59–56
red ER, yellow MR	$< 5.6 > 5.6$	c = 56–56
red ER, indet MR, yellow BAN	$< 5.6 > 4.8$	d = 56–48
red MR, yellow BAN	$< 5.2 > 4.8$	e = 52–48
red MR, indet BAN, green to blue BCG	$< 5.2 > 4.4$	f = 52–44
red BAN, green BCG	$< 4.4 > 4.4$	g = 44–44
red BAN, indet BCG, green to blue BPB	$< 4.4 > 4.0$	h = 44–40
yellow BCG, green to blue BPB	$< 4.0 > 4.0$	i = 40–40
yellow BPB	< 3.4	k = < 34
Wider Ranges (indefinite)		
yellow BCP, green to blue BPB	$< 5.9 > 4.0$	X = 58–40
red ER, green to blue BPB	$< 5.6 > 4.0$	Y = 56–40
red MR, green to blue BPB	$< 5.2 > 4.0$	Z = 52–40
Higher Ranges		
yellow BTB, indet BCP, yellow ER	$< 6.2 > 5.9$	B = 62–59
yellow BTB, blue to purple BCP	$< 6.2 > 6.2$	C = 62–62
yellow PR, indet BTB, and blue to purple BCP	$< 6.8 > 6.2$	D = 68–62
yellow PR, green to blue BTB	$< 6.8 > 6.4$	E = 68–64

Small's Range Indicator Method

The early estimations of pH in plant cells, by means of indicators, involved comparisons of the various tints of a dichroic fluid with the indicators in different concentrations. In 1921 Small devised his Range Indicator Method and used it with his collaborators for five years, to make sure it was a practical method, before publishing it in detail with some of the results in 1926 (Small, 1926); Rea and Small, 1926, 1927 (a) and (b); Small, Martin, and Rea, 1926; Martin, 1927 (a) and (b); see also Small, 1929). Since then the principle of Small's R. I. M. has been widely recognised as sound in practice and logical in theory.

This principle recognises the various errors which may invalidate interpretations of tints of indicators used on living cells; and in the R. I. M. method colours, and not tints, are taken as the only valid indications. Each indicator is used to define the upper or lower limit of a range of pH within which the pH of the observed material must lie; and by using a number of indicators with colour changes above and below this range the R. I. M. pH range is checked as to its upper and lower limits. The precise pH of the material may not be obtained, but the range within which the precise pH is situated is defined without any doubt. When with two indicators one indicates below a certain pH and the other indicates above the same pH, the true value is taken as being the pH indicated by both indicators taken together within about .05 of a pH unit, *e.g.* <5.9>5.9 = 59–59 in Tables II and III, similarly the points 40–40, 44–44, 56–56, 62–62 are obtained sometimes, but usually the result is a wider range such as 59–56, 52–44, or the common wide range 52–40.

The indicators used in SMALL's R. I. M. are given in Table I, and discussed later.

Table III. *Outline of new Notation for R. I. M.*

As a working summary for the interpretation of R. I. M. data and for the easier comprehension of extensive series of results the following diagram has been found useful:—

pH ranges, R. I. M., more acid to less acid

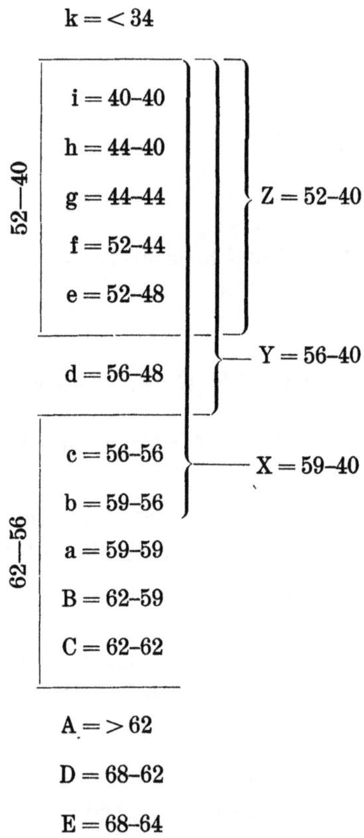

The pH ranges indicated by the colours obtained using the R. I. M. series of indicators are listed in Table II, and are discussed later.

Most of the results published by SMALL and his collaborators were tabulated extensively under alphabetical symbols for the various pH ranges. This alphabetical notation has been found to be rather difficult to follow unless one is completely familiar with the letters and equivalents. In order to make any detailed or connected account more readily comprehended, a new notation is given in Table III, as 52–40 for Z which stands for "a precise pH value within the range below pH 5.2 and above pH 4.0": note that 52–40 does *not* mean a pH value which v a r i e s f r o m pH 5.2 to pH 4.0.

The diagrammatic form of Table III distinguishes two main ranges for plant cell-saps, namely 62–56 which is the common normal range and 52–40 a more acid range that occurs in the epidermis in many stems and in all or most cells of plants belonging to the 'acid-families..' It should be noted that this more acid range extends over only 1.2 pH units, while the so-called normal range 62–56 occupies only 0.6 of a pH unit, and that there is the gap between 56–56 and 52–40 which is covered as a minimum range 56–48 when ER is red < 5.6, MR is indeterminate and BAN is yellow > 48. This intermediate range is also covered by wider ranges such as 56–44 and 56–40. The narrowness of the two common ranges suggests that there is a considerable stability at two different levels of metabolism in plant cells, one where an excess of free organic acids keeps the pH within the 52–40 range, and the other level at which the sap is nearly neutral except for the acidifying effect of its carbon dioxide content and possibly traces of organic acids.

Cell Sap of *Helianthus annuus*

Sunflower cell-sap

As one example of the convenience of the notation using round numbers for what may be called Sörensen units, the summary given by Martin (1927 a) of her detailed results concerning the pH of cell-sap at all stages and in all parts of the sunflower *Helianthus annuus* may be re-written as follows:—

The cell-sap in the epidermal cells is relatively acid throughout, pH in the 44–40 range, except when the cells appear to be injured and then the sap is < 34. This 44–40 outermost layer covers the embryo; the cotyledons, stem and root of the seedling; the stem, petiole and leaf up to the expansion of the second foliage leaf. At later stages of growth, the epidermal cells of the lower surface of the third (and most younger leaves) become less acid, 52–48, or even 59–56 or 59–59, while all other epidermal cells remain 44–40.

On the stem and under surface of leaves of sunflower there are two types of hairs; the common type is one cell thick from tip to base with pH in range 52–48; the second type has a thick multicellular base and gives a deep blue with thymol blue (TB) indicating pH > 9, as well as alkaline colours with BPB to PR, and with neutral red and cresol red; but is not red with phenolphthalein indicating pH < 10. These hairs are the most alkaline objects we have found in flowering plants.

In the sunflower root the piliferous layer and root-hairs are acid, usually within the 44–40 pH range at all stages of growth; and so is the root-cap, root-tip and dermatogen of the radicle in the seed and young seedling. The plerome and inner periblem of the radicle is in the 52–48 pH range. The exodermis in the older, upper and middle, parts of the root is in the

44–40 pH range in all stages after its differentiation, but the acidity decreases into the 52-48 range near the tip of the root from an early stage until the expansion of the fifth leaf; then the acidity of the exodermal sap decreases still further, into the 59–56 range, with some cells remaining in the 52–48 range.

The cell-sap of the cortex and pith in aerial parts of the sunflower plant is usually in the 59–56 range, with some cells definitely 59–59 especially in mature plants. In subterranean parts the cortex and pith are more acid, usually 52–48, but sometimes the higher range of 59–56, occurs.

The endodermal cell-sap above ground is 59–56, sometimes 59–59, while below ground the same zone of cells is in the range 44–40, sometimes 52–48, and occasionally very acid, < 34.

The pericycle underground in most cases in less acid (52–48) than the endodermis (44–40), but in a few cases 44–40 and even < 34 were recorded. Above soil-level the sap of the thin-walled pericycle cells corresponds in pH with that of cortex and pith, at 59–56 with 59–59 sometimes.

The walls of the mature pericyclic fibres and of the mature xylem fibres and vessels are among the most acid parts of the sunflower and of most other plants. In this sunflower study, during the process of lignification a progressive increase of acidity was noted; virages corresponding to 52–48 passing with time to 44–40, and finally to < 34 occurred in fibres, but vessel-walls remained in the 44–40 range throughout the plant.

The phloem is not alkaline, but it is less acid than the xylem parenchyma (44–40) being usually 52–48 in the younger stages of growth and passing to 59–56 or 59–59 in mature plants. The callus found on the sieve-plates in the autumn gives colours which indicate 62–62. This might yield a bluish colour with some samples of litmus and so give rise to the fable that phloem is an alkaline tissue.

Cambial cells do not differ from neighbouring phloem or cortical cells. In older stems some phellogen cells are interesting: the outer cells cut off from this meristem may be in the 52–48 range while the phellogen cells remain in the 44–40 range.

The mesophyll in cotyledons and in young foliage leaves have cell-sap mostly in the 52–48 range, but some young foliage leaves show 59–56, and this less acid range is common for the mesophyll in mature plants.

Young, developing, flower-heads have their sap pH in the 52--48 range, but parts as they mature become differentiated in their cell-sap pH values. The style becomes more acid at least to 44–40 and sometimes to < 34; while the anthers become less acid, maturing into the 59–56 range, but with pollen-grains in the 52–48 range. The ovary and its single ovule remain in the 52–48 range at all stages. The corolla, when developed, has the usual acid (44–40) epidermis, but the cell-sap of the receptacle remains in the 52–48 range.

This summary of sunflower cell-sap pH values is rather more than just an example of the numerical notation for the R. I. M. It serves as an introduction to the extensive R. I. M. results given later, and also as a standard which may be described as normal, although it includes an all-over acid epidermis which definitely does not occur in all plants, and also some alkaline epidermal hairs which appear to be quite exceptional. It should be noted that the pH range 52–48, as a part of the 52–40 range, is quite a common range for cell-sap in the sunflower (*Helianthus annuus*).

This summary, together with Tables I, II, III, is introduced at this early stage so that it may serve as a basis for comments on the significance of early, and later, results of observations which include comparisons of tints given by indicators in concentrations and thicknesses of layers of fluids varying to an undetermined extent.

II. The Modern Outlook on pH and the R. I. M.

In order to understand the changes in the significance of earlier records of pH values for plant cells, it is very necessary to have some knowledge and appreciation of the modern changed outlook on the validity and degree of precision of estimations of pH under controlled conditions *in vitro*. When these values are recognised as being subject to an experimental variation covering 0.04 in the estimation of pH in reasonably pure simple mixtures of chemical substances, it becomes doubtful whether the precision of estimations of the pH of cell contents can ever attain an accuracy of more than ± 0.05 of a pH unit.

In addition to this lesser degree of precision, there has been a change in the official views concerning hydrogen ion concentration and hydrogen ion activity. Both points are covered in the pamphlet on "The pH Scale," published as "British Standard 1647 : 1950" by the British Standards Institution, 24/28 Victoria Street, Westminster, London, S.W. 1, to which acknowledgments are due for permission to use quotations. Both aspects are also included in the American Letter Circular LC 993 of the U. S. National Bureau of Standards (Department of Commerce, Washington 25, D. C., U. S. A.) from which quotations are given with all due acknowledgments of their source.

It is now recognised officially that hydrogen ion concentration as such is not measurable, and that what is measurable with a limited degree of accuracy is the total activity of hydrogen ions in relation to the activities of other ions in the same solution. This substitution of activity for concentration was advocated by Lewis (1912) and more or less accepted by Sörensen, also by Michaelis who (1926, p. 128) wrote that "the true concentration of H-ions is, on the other hand, entirely unknown. We have no direct method of measuring it, and probably it is of no great interest to us, for the effectiveness of the H-ions appears to depend only upon the *a* (H^+), the H-ion activity." Ricci (1952, p. 7) seems to agree and gives as "the most fundamental equilibrium constant"—

$$K_w = a_H + a_{OH^-} / (a_{H_2O})^h = \mathbf{H\,(OH)} / (a_{H_2O})^h$$

where all terms refer to activities.

Since the activity coefficients for various ions are taken as valued at 1 for infinitely dilute solutions and decrease in value to various extents with increase in concentration, the measurement of total hydrogen ion activity can be only approximate in practice, although RICCI (1952) using various *general* mathematical assumptions solves his mathematical problems with mathematical precision (cp. BATES 1954).

The modern official standards for the estimation of pH are definitely practical. The scope of the British Standard 1647 : 1950 is limited to "aqueous solutions at temperatures between 0° C. and 60° C. The standard is intended to be accurate within ± 0.005." But it is recognised that this degree of accuracy is "rarely obtainable with the glass electrode." In clause 5 it is stated that "in the restricted range of dilute aqueous solutions (not exceeding one-tenth molar) ... (pH between 2 and 12); the definition is such that—

$$pH = -\log_{10} C_H f_1 \pm 0.02$$

where C_H denotes the concentration of hydrogen ion in g. ions/litre, and f denotes the mean activity coefficient of a typical uni-univalent electrolyte in the solution." "There are at present only about half a dozen suitable solutions the pH values of which, determined according to this British Standard, are reliably known to an accuracy of ± 0.005. The American LC 933 gives some of these solutions as standards, in addition to the British primary standard, which has *by definition* a pH value of 4.000: this standard is "a one-twentieth molar solution of pure potassium hydrogen phthalate," ($HOOC \cdot C_6H_4COOK$). "The pH of this solution shall be defined as having the value 4 exactly at 15° C. At any other temperature t° C. between 0° C. and 60° C., its pH shall be defined by the formula:—

$$pH = 4.000 + \frac{1}{2}\left(\frac{t-15}{100}\right)^2 \text{"}.$$

The standard solutions given in LC 993, with the variation of pH over various temperature ranges are as follows:—

0.05 M potassium hydrogen phthalate, pH 4.01–4.10 (0°–60° C.);

0.01 M potassium tetroxalate, pH 2.14–2.16 (0°–38° C.);

saturated solution of potassium hydrogen tartrate, pH 3.56–3.55 (25°–38° C.)

0.025 M NaH succinate and Na_2 succinate dissolved in the same water (0.05 M succinate), pH 5.46–5.41 (0°–38° C.);

0.05 M phosphate, as 0.025 M KH_2PO_4 + 0.025 M Na_2HPO_4 dissolved in the same water (0.05 M phosphate), pH 6.98–6.84 (0°–60° C.);

0.01 M borate, $Na_2B_4O_7 \cdot 10\,H_2O$, pH 9.46–8.96 (0°–60° C.);

0.05 M carbonate, as 0.025 M $NaHCO_3$ + 0.025 M Na_2CO_3 dissolved in the same water, pH 10.32–9.91 (0°–38° C.);

0.01 M trisodium phosphate, pH 11.72–11.38 (25°–38° C.).

Thus the variation of pH in the temperature ranges given is not more than 0.10 of a pH unit for phthalate, tetroxalate, saturated tartrate, succinate, and trisodium phosphate, and is 0.14 for phosphate, 0.50 for borate, and 0.41 for carbonate. The wide variation of carbonate pH in the natural temperature range of 0^0–38^0 C. may be of some significance in both estimations of natural pH values and in some slightly buffered plant fluids.

LC 993 also gives some useful data concerning (a) the pH changes on dilution of these standard solutions with an equal quantity of water and (b) the approximate buffer index value (dB/dpH) for each solution around its natural pH value.

One of the main objectives in the adoption of a somewhat arbitrary practical standard for pH estimations is *universal* reproducibility in relation to standard preparations, pure solutions and mixtures. Some degree of reproducibility may be attainable with the contents of plant cells, but with living material which is sensitive to environmental conditions and varies within itself as well as varying with its source, the degree of reproducibility of pH values may well be not even to the first decimal of the pH number.

This brings us back to the pH ranges obtained by means of SMALL's R. I. M. There are five ranges which are around (± 0.05) a given pH value, namely 40–40, 44–44, 56–56, 59–59, and 62–62, and other ranges extend from 0.3 to 1.9 of a pH unit. Details of this method have been published several times (SMALL, 1926, 1928, 1929, 1946) but precautions to be taken in the application of this method are very important for obtaining correct observations and for drawing the right conclusions from the observed colours.

R. I. M. Indicators

The indicators specified in Table I should all be *used* as aqueous non-alcoholic solutions but ethyl red (ER) in the red form does not keep well without about 50 per cent alcohol and such a strength of alcohol is rapidly toxic. This difficulty can be met by storing an aqueous solution of ER in the alkaline or yellow form since it is the red acid form which separates out from aqueous solutions as crystals. Some of the stored solution can be titrated with N/20 HCl to the orange tint around its change-zone for use with plant cells. In fact all the other indicators should, immediately before use, be similarly brought into their change-zone by titration with N/20 HCl or N/20 NaOH, so that they are easily thrown by the plant fluid to one side or the other of their change-zone, unless the pH value of the plant-fluid happens to lie within the change-zone of the indicator, in which case the R. I. M. record for that particular indicator will be indeterminate or "indet." Benzene azo-α-naphthylamine (BAN) is another indicator which does not dissolve readily in water alone: for use in the R. I. M. it should be dissolved in absolute alcohol (0.01 g. in 2 mils) acidified with two drops of normal HCl, and diluted with distilled water to 100 mils. This solution should be red; and for immediate use in the R. I. M. the acidity should be reduced by titration with N/20 NaOH until the addition of one drop of that alkali leads to the formation of the yellow precipitating form of BAN; precipitation is then counteracted by the addition of one drop of N/20 HCl to 100 mils of indicator solution (COOKE, 1952).

All the other R. I. M. indicators can now be obtained by specifying "water-soluble" when ordering the dry solid materials. Commercial solutions of indicators should not be used for the R. I. M., as they may contain indefinite percentages of alcohol.

The concentrations of indicators are 0.02 per cent for MR and ER, and 0.04 per cent for all the others except BAN which is only 0.01 per cent.

R. I. M. Precautions

As one result of teaching the use of Small's Range Indicator Method to various kinds of students over a period of over thirty years, Small can now indicate a number of precautions in the absence of which R. I. M. observations may be inconsistent within a particular series, or difficult to interpret, or quite aleatory.

(1) The observer should be tested for colour blindness. There has been in my experience a larger than normal proportion of men students definitely colour blind, and most of the successful workers with pH indicators have been women.

(2) The indicators should be stored as aqueous solutions in amber glass bottles. For immediate use small quantities (10–20 mils) should be transferred to white glass bottles fitted with pipettes, and this quantity of indicator solution should be titrated judiciously with N/20 hydrochloric acid or sodium hydroxide so that the indicator is about the middle of its change-zone which extends over about 1.6 of a pH unit.

(3) The material to be examined should be fresh and used uncut, or if sections are cut they should be washed with freshly distilled water (not stored for more than two days). The ordinary 'blow' wash-bottle should not be used, since the carbon dioxide of the breath acidifies the water rapidly. Initial observations of colours should be made under low power magnifications, and a coverslip should not be placed upon the section as the respiratory carbon dioxide of the plant tissues may induce acidification, so also may the breath of the observer unless care is taken not to breathe upon the exposed section down the side of the microscope. In cases where acidification does not take place under a coverslip closer observation at higher magnifications becomes possible.

(4) The duration of the period of immersion of the material in the indicator solutions should be reduced to a minimum consistent with obtaining unequivocal colour indications: this time may vary from one minute to twelve hours, and is a matter for observation with any given material.

Used with all these precautions, and interpreting the series of colours obtained with an allowance for the inherent experimental error of any indicator method ($\pm$ 0.1), the R. I. M. can be used with some hope of getting natural pH values. The variations which occur from one tissue to another, and in the same tissue from one time to another, are often larger than the ranges defined by R. I. M. colours, so that with at least one of the series of indicators there is normally an obvious differentiation in the pH values for different tissues in the same section or sometimes for the same tissue at different seasons.

R. I. M. and other Methods

The use of the R. I. M. in accordance with the detailed procedure, in the estimation of the pH of plant cells, avoids entirely the large general errors of expressing mixed cell-saps, and to a considerable degree the R. I. M. avoids also the following errors known for other methods:—

(a) Hydrogen Electrode Errors—due to carbon dioxide dilution or elimination, reduction, oxidation, electrode poisoning, etc.;

(b) Quinhydrone Electrode Errors—due to protein, phosphate, glucose, tannin, and carbon dioxide dilution;

(c) Micro-Electrode Errors—due to above factors, especially electrode poisoning, together with membrane formation and other small scale disturbances;

(d) Comparator and Capillator Indicator Methods—due to self-colour, dilute solutions, salts, proteins, lipoids, temperature, chemical changes, etc.;

(e) Errors of Other Indicator Methods—due to adsorption, tint-comparisons involving concentration or localisation of indicators, solubility of indicators, self-colour of cells, displacement of carbon dioxide, lipoids, dielectric constant, toxicity in living cells. The toxicity of some indicators has been proved for seedlings *grown in* solutions of indicators, but used as in the R. I. M. for short time immersion any toxic indicator would be readily detected by plasmolysing and deplasmolysing the cells examined.

Most of the R. I. M. indicators are not toxic within the time-limits for observations of the colours of absorbed indicators. Sections and other parts of plants can normally be left for hours and become distinctly coloured long before any toxic effect can be detected by the plasmolysis-deplasmolysis test.

R. I. M. Limitations

The results obtained using the R. I. M. with the procedure specified above are subject to the following limitations:—

(1) The estimation of a pH range, rather than a point to the first decimal $\pm .05$, may give the same range for two or more kinds of cells which may be different in their actual pH values, if the differences happen to lie within the range determined. This error, or rather lack of differentiation, is minimised, but not completely avoided, by the use of more indicators; *e.g.* by the breaking up of the common 52–40 range into the ranges 52–48, 52–44, 44–44, 44–40, 40–40, by the use of BAN and BCG between the red for MR (<52) and the green or blue for BPB (>40).

(2) Two of the more acidic indicators, BPB and BAN are toxic to varying degrees, and symptoms of toxicity should be noted if they occur, but these indicators are rapidly absorbed and show distinct colours with many plant materials without any indication of toxic effects when prepared as aqueous non-alcoholic solutions like all other R. I. M. indicators. Even when toxic effects result in indeterminate colour indications with these two indicators the yellow or green-to-blue colours of BCG serve as a check for >44 or <40.

Perhaps the best test, as to whether the R. I. M. results are somewhere near the truth for natural pH values, is the appearance of definite colour differentiation between one cell or part of one cell and another cell or cell-part. If all the material takes up the same colour as the indicator solution, with all the different indicators, there is something wrong with the technique as being applied. Alcoholic indicators or abnormally rapid toxicity may give such results, and so may colour-blindness on the part of the observer.

With these limitations, the R. I. M. has been proved to be an effective reliable method over a wide range of material and of observers during the last thirty years; and SMALL's R. I. M. with the particular indicator solutions specified, used with the procedure as outlined above, is still part of the modern outlook on pH, with the newer latitude in accuracy for *in vitro* estimations approaching more nearly to the R. I. M. pH ranges.

III. pH and Natural Indicators

The earliest observations of the pH of plant cells were made on flowers and other parts of plants in which anthocyans act as natural internal indicators of degrees of acidity. Being already naturally inside the cells these indicators have great importance in so far as they can, in at least some experimental cases, be used in connection with other physiological phenomena such as the ready penetration of carbon dioxide and ammonia into living cells. The actual natural pigmentation of flower petals is another matter, since the natural variation in colour may be complicated by the presence of two or more forms of anthocyanins, by co-pigmentation with flavones or tannin, and by the particular colloidal condition of the pigments, as well as by the actual concentration of colouring matter or relative proportions of two or more anthocyans in the cells.

All these complications are stressed by BLANK (1947) to such an extent that to that author it seems to be more advisable to theorise about colloidality or co-pigmentation in unproven cases, than to apply the rule of economy of hypotheses and accept a correlation of colour changes *in vivo* with demonstrated colour changes in buffered solutions of known pH values as evidence favouring pH as an active and important direct factor governing within limits the colour of any anthocyanin or combination of anthocyanins. Colloidality and co-pigmentation have been proved to be effective factors, sometimes controlling differences of colour within the same narrow range of pH values, but this does not invalidate the pH control of colour in other cases, especially those in which rapid and reversible colour changes have been produced experimentally, see SMITH (1923) and others as presented below. Co-pigmentation may include the formation of tannates of anthocyanins as oxonium salts of the amphoteric pigments.

In addition to floral anthocyans, quite a number of plant products have been used, even from Roman times, as *in vitro* indicators of acidity or alkalinity, such as red wine, syrup of violets (BOYLE, 1664) litmus, cudbear, orchil, turmeric, alkannin, logwood, mimosa flower extract, and parietin (see also CLARK, 1928, p. 86), but here we are concerned with these and others only in so far as they occur dissolved *within* plant cells and by their colours indicate the pH of plant cell sap.

Before the pH scale was invented, SCHWARZ (1892) extracted the colouring matter of some flowers and showed that the change from red to blue during the anthesis of flowers of *Pulmonaria, Anchusa* and *Lathyrus* corresponded to a similar colour change with decrease of acidity *in vitro*. After the pH notation for degrees of acidity was introduced in 1909, more accurate observations of cell-sap pH became possible, using natural indicators and checking the natural colours against colours obtained with these anthocyans *in vitro* in buffered solutions at various pH values. WALBUM (1913) introduced the use of an extract of red cabbage as a pH indicator for use *in vitro*, but this was more an extension of the well established practice of using plant products as pH indicators, rather than observations on natural cell pH values. HAAS (1916) gives pH 6–7 as the natural range for red cabbage leaves. WATSON (1913) described the use of the juice of the blueberry as a pH indicator.

SCHLEY (1913) described a natural indicator in the testa of ripe seeds of *Vicia faba* as grey and opalescent in solution, changed by alkali to yellow and then to "the color of the testa of the ripe *Vicia faba* seed." The writer has observed the acid-green, alkaline-brown indicator of the broad bean seed, also the acid-colourless, alkaline-yellow indicator of the maize fruit. SCHLEY (1920) used the bean-testa indicator in further work on chemical changes during geotropic response.

MCCLENDON (1914) observed another indicator in the nectar glands of *Vicia faba* which showed red-acid, blue-alkaline, and gave rise to colour changes during the functional activity of the glands. No pH values were given. He also noted the natural indicator of the red cabbage (*Brassica oleracea* var. *capitata* f. *rubra*).

WILLSTÄTTER (1914) recorded that the same anthocyanin (cyanin) with different pH values for the cell-sap gave redness in rose petals at pH 5.5 and blueness in the corolla of the cornflower at pH 7.2. He also recorded that in buffered solutions the rose petals go blue at pH 7.2, and the cornflower corollas become red at pH 5.5. A simple experiment on similar lines consists of cupping a purple field scabious flower-head in two hands and blowing alkaline tobacco smoke (or ammonia vapour) upon the blossom which soon turns a brilliant green. The cornflower story was complicated by ROBINSON, see BLANK (1947), who quotes (*op. cit.*, p. 246–247) cyanin as red at pH 3.0 or less (in the form of its oxonium salt); violet at pH 8.5 (in the form of the colour base for which a quinoid structure is given); and blue at pH 11.0 (for which the sodium salt of the colour-base is illustrated). Anthocyanins are amphoteric, and the red form is more or less a chloride or other salt of the colour-base while the blue form of cyanin has the colour-base as an acid radicle for sodium or potassium.

BLANK then (*op. cit.*, p. 255–256) quotes R. and G. M. ROBINSON as having "observed that the cell-sap of cornflower colours blue litmus red and even has a more acid reaction than the cell sap of the red rose." Then BLANK (*loc. cit.*) proceeds to theorise about a complex, and brings in a lyophilic colloid in the cell sap, and mentions xylan·and other polysaccharides in extracts of cornflowers. He quotes the ROBINSONS as suggesting "that the cyanin in the blue cornflower cell sap is blue because it is absorbed on colloidal particles which stabilize the anions at a pH of 4.9," but all this, while it may apply to the natural cornflower blue, neither contradicts nor explains WILLSTÄTTER's experimental changes of colour in buffered solutions.

HAAS (1916) investigated the acidity of plant cells as shown by natural indicators with their natural colours checked *in vitro* by means of buffered solutions at various pH values. He found that a considerable change of pH, up to as much as from pH 3 to pH 7, may occur in a cell as it dies. Using natural colours, he recorded pH 2–3 for radish root, pH 3 for red beet root, pH 6–7 for red cabbage leaves; using anthocyans of petals he found natural cell-sap pH values as follows:—

Viola tricolor 4, *V. odorata* 5; *Primula obconica* 6, *P. chinensis* 6–7, *Hyacinthus* blue 4.5, red 7; *Cichorium intybus* 3, *Scilla* 4.5, *Browallia speciosa* 6, *Pelargonium* 8.

CROZIER (1916) described natural indicators in various fruits, including one in the red berries of *Lantana involucrata* which changes from its natural red to green then yellow above pH 6.6, and another indicator in the fruits of *Randia aculeata* showing red when acid but green above pH 8.5.

JACOBS (1922) used red *Rhododendron* blossoms in an investigation of the ready penetration of NH_4 into cells; and found that this anthocyan which is red below pH 7, changes through violet to blue at pH 8 and above. The differential penetration of NH_4 from even acidic external solutions produces an internal decrease of acidity, and in the same way a differential ready penetration of carbon dioxide from an alkaline external solution of $CO_2 + NaHCO_3$ increases internal cell-acidity.

ATKINS (1923) recorded plants of *Hydrangea* as pink or blue on soils of various reactions, usually pink when the soil-pH was above 7.5 and blue when it was 6.0 or 5.75, but with both colours at 5.9, 6.2, and 7.3. He found, however, that the anthocyan in this case is *not* a pH indicator since the expressed saps from the petals (by a drop comparison method) was of pH 4.0–4.2 in both pink and in blue flowers. For this outstanding exception to the control of reds and blues by pH ATKINS suggested that the blue colour required iron for its development, but he wrote (*op. cit.*) that "It is possible, however, that the aluminium, as well as the iron, may form a blue complex with the anthocyanin, which is pink in the absence of excess of these salts." CHENERY (1937) found in blue *Hydrangea* flowers an Al : Fe ratio of nearly 16 : 1, while ALLEN (1943) found the Al : Fe ratio about 1 : 1 with the soil-pH above 5.5 and the flowers pink to pinkish-blue; but Al : Fe as 2375–897 p.p.m. Al to 516–310 p.p.m. Fe with the soil-pH at 4.56–5.13 and flowers blue. So that in this case, as in many other cases, the pH value is only one link in a chain of processes. The pH effect upon the difference in *Hydrangea* flower colour is not direct: increased soil-acidity results in an increase of both iron and aluminium in the coloured petals. ALLEN changed petals from pink to blue on one branch of a pink-flowered plant by spraying it with a solution of aluminium ammonium citrate. Other experimental treatments carried out by ALLEN leave very little doubt that the *Hydrangea* requires freely available aluminium in its petals before it can produce blue flowers, and ATKINS' 1923 suggestion of an aluminium complex with the anthocyanin seems to be supported by this later work, although the relation between soil-acidity and flower-colour is much more complex than would appear from this brief summary. The originals should be consulted for their many details.

Returning to natural indicators which are not complicated by aluminium requirements, we have SMITH's work on petal colours, which began in 1923 with observations on flowers of *Ipomaea learii* where the anthocyan is red in the bud (pH 6), blue during anthesis (pH 7.8) and red again when the flower is fading (pH 6). In accordance with the now well known ready penetration of very weak acids, water saturated with carbon dioxide (pH 5) was found to change the virage of discs of these petals from blue to red in a reversible fashion, although "strong acids" at pH 5 were without influence on the colour except at cut edges of the discs. These blue-red-blue changes with carbon dioxide and with plain water were rapid and readily reversible, as they would be with a simple artificial indicator. SMITH continued publication later (1931, 1933) on work involving anthocyans, but meanwhile BUXTON and DARBISHIRE published two extensive investigations concerning pH in relation to flowers (1929a, 1929b).

The first reference (1929a) concerns the pH of petal cell-sap and includes separate accounts of observations made—(I) on dried petals, collected during the summer, partially dried at 25° C. and completely dried at room temperature in a desiccator; then ground to a powder and stored until tested for the pH values which developed in the filtrate when 1 gramme of petal powder was macerated overnight in 200 mils distilled water at pH 6.8–7.0, the filtrate kept for three months with a little thymol added to prevent the growth of micro-organisms, and tested at weekly intervals with a comparator: (II) on fresh petals, 10 g. "powdered up in a mortar," "filtered with the help of a vacuum pump," diluted 1 to 25 with neutral distilled water, and tested with a comparator immediately and also at weekly intervals for three months. The end results in all cases, after three months, obviously have little or no bearing upon the natural pH values of the original living material, but the initial pH values with the second method do refer to pH numbers for mixed saps. In some cases undiluted extracts from fresh *white* flowers were tested by the capillator method. and these results should approximate closely to natural values for mixed sap. All these were found to be distinctly acid, and may be quoted as of some value in connection with the present subject of pH of plant cells, although being from *white* petals, they have little connection with natural indicators: they are—rhododendron 4.0, peony and Shirley poppy 4.5, lupin 4.6, rose 4.7, aquilegia 4.8, primrose and *Iris kaempferi* 5.0, and *Campanula* 5.5. The average initial pH values of the diluted saps (1 to 25) as given in graphs was about 6.2–6.6, or 6.4 as given in the later paper.

The second reference, Buxton and Darbishire (1929 b) concerns the behaviour of "anthocyanins" at various pH values. Two grammes of petals were pulped with pumice in a mortar; the pulp was divided into nine equal parts and placed in nine different tubes of buffer solution at pH 3, 4, 5, 6, 7, 8, 9, 10, 11. As controls small pieces of petals were kept in nine corresponding but smaller tubes of buffer solutions. The larger tubes were well shaken and the colours shown were recorded as soon as the larger particles had subsided: except with colours at pH 3, fading began in less than one hour and was more or less complete within 24 hours. On the basis of this series of observations with a technique which seems quite in accordance with generally accepted principles, Buxton and Darbishire tabulated the colours given by the anthocyanins (as oxonium salts, colour bases or as potassium salts, with possibly colloidal and co-pigmentation complications) and flavones as producers of a yellow colour component particularly prominent in the range pH 8–11, as a constituent yielding green with pale anthocyan blues which fade quickly leaving a deep yellow after a few hours. A colour plate of six selected examples is a valuable addition to the Tables of colours from pH 3 to pH 11 given for 13 blue flowers, 10 red flowers, and 18 intermediates, 6 more blue than red, 4 more red than blue, 3 'acid,' 4 white, and blue pimpernel as "aberrant."

Species are mentioned in these Tables only occasionally, so the data are not very precise as to the source of the anthocyans, but a summary is of some considerable interest. Using the following abbrevations for colour

Table IV.

(Table I ex Buxton and Darbishire, 1929 b.)

Blue Flowers

pH	3	4	5	6	7	8	9	10	11
Aconitum	pk	pkv	v	vB	B	g	g	g	g
Aquilegia	R	vpk	vB	B	B	gB	Bg	g	g
Campanula	R	vpk	v	vB	B	gB	g	g	g
Clematis	pk	pk	v	v	B	—	g	g	g
Delphinium	R	v	B	B	B	gB	g	g	g
Gentiana acaulis	pk	v	vB	B	gB	Bg	g	g	g
Gentiana asclepiadea	pk	v	vB	B	gB	Bg	g	g	g
Iris kaempferi	R	pk	vpk	v	vB	B	B	gB	gB
Lathyrus (purple)	R	pk	v	v	B	B	gB	Bg	g
Lathyrus (blue)	pk	pk	v	v	B	gB	Bg	g	g
Linum	R	pk	pkv	v	B	gB	Bg	g	g
Primrose	R	pk	v	v	B	B	gB	Bg	g
Salvia (patens)	pk	B	B	B	B	B	gB	g	g

(Table II ex BUXTON and DARBISHIRE, 1929 b.)

Clear Red Flowers

pH	3	4	5	6	7	8	9	10	11
Begonia	VM	VMpk	pk	pk	pu	—	brpu	pu	pu
Dahlia	VM	o	o	oR	oR	R	R	R	R
Lathyrus	VM	VMpk	pk	pk	pk	pk	pu	pu	pu
Papaver orientalis	VM	VMpk	pkR	R	puR	puR	Rpu	Rpu	Rpu
Papaver rhoeas	VM	VM	pk	pk	obr	obr	Rpu	Rpu	Rpu
Rose	VM	pk	pk	pk	pu	R	R	pk	br
Salvia (splendens)	VM	VMpk	VMpk	R	puR	puR	pu	pu	pu
Tropaeolum	VM	VM	RVM	R	puR	puR	puR	puR	pu
Acid Pelargonium	VM	pk	pk	pk	v	v	pu	pu	pu
Acid Rhododendron	VM	R	pk	v	v	puR	puR	puR	br

(Table III ex BUXTON and DARBISHIRE, 1929 b.)

Intermediate Flowers

Sub-group A. More blue than red.

pH	3	4	5	6	7	8	9	10	11
Anchusa	R	pk	v	B	B	pu	pu	Bg	g
Digitalis	R	pk	pk	vpk	vB	pk	brpk	g	g
Iris siberica	R	vpk	pkv	v	B	B	pu	g	g
Lupin	R	vpk	v	v	B	Bbr	br	gbr	g
Salvia farinosa	vpk	bry	ybr	vbr	vbr	v	Bg	g	g
Viola	R	pk	v	vB	B	puB	pu	g	g

Sub-group B. More red than blue.

Dahlia	R	pk	pk	vR	v	R	Rbr	gbr	brg
Linum	VM	Rpk	v	v	B	puR	puR	pu	pu
Primrose	R	R	pk	v	v	v	pu	pu	g
Rose	VM	pk	pk	vpk	v	R	puR	g	g

Sub-group C. Acid.

Paeony	R	pk	pk	pk	v	v	B	gB	g
Pelargonium	VM	pk	pk	v	v	v	B	gB	g
Rhododendron	R	pk	vpk	vpk	v	vB	B	gB	g

White Flowers

pH	3	4	5	6	7	8	9	10	11
Dahlia	w	w	w	py	py	dy	dy	dy	dy
Iris kaempferi	w	w	w	w	w	w	py	py	py
Rhododendron	w	w	w	py	py	py	dy	dy	dy
Rose	w	w	w	py	py	py	dy	dy	dy

Aberrant

Anagallis (blue)	R	v	B	pu	g	g	pu	pu	pu

names:— B = blue, R = red, VM = vermilion, pk = pink, v = violet, g = green, o = orange, br = brown, pu = purple, w = white, py = pale yellow, dy = deep yellow; and various combinations of these, the Tables can be contracted with very little modification as follows:—

The colours recorded in Table IV (Tables I–III) are for pulped petals in aliquot portions in 10 mils of buffer solutions at the various pH values, but pieces of petals, not mashed, in smaller tubes of a similar series of buffers, "served very well as controls," so the unstated implication is that whole pieces of petals gave colours similar to those given by pulped petals. Taking this as true, these Tables demonstrate that there are at least two types of anthocyanins which are blue (B) around neutrality and pink (pk) or red (R) at pH 3, with variations through violet from pH 4 to pH 6. The colours at higher pH values, 8 to 11, are experimental artefacts which have little or nothing to do with *natural* colours of flowers.

The possible significance of these records may be considered in the light of later work and data gleaned from BLANK's extensive survey of the literature about anthocyans (1947). BLANK (*op. cit.*, p. 242) describes anthocyanins as soluble in water, as glycosides of anthocyanidins which are the sugar-free or aglycones of anthocyanins. Anthocyanidins are, on p. 246, described as "not soluble in water," and (p. 242) as "observed in plants only in rare cases." Anthocyanins are amphoteric, and their salts in which they act as bases *e. g.* chlorides or tannates are called oxonium salts to distinguish these from the salts formed in which the anthocyanin acts as an acid radicle for a base such as potassium or sodium. In the isoelectric zone for these amphoteric pigments the colour-base is free. The oxonium salts of anthocyanin are reds of various shades; the colour-bases are violet for cyanin at least, while sodium or potassium salts of the colour-base are usually blue or violet. The details of the constitution of anthocyans were expounded by SCOTT-MONCRIEFF (1932). Three types of aglycones were given:— Pelargonidin with an OH at 4′; cyanidin with two OH, one at 4′ and one at 3′; delphinidin with three OH at 3′, 4′ and 5′. These OH groups

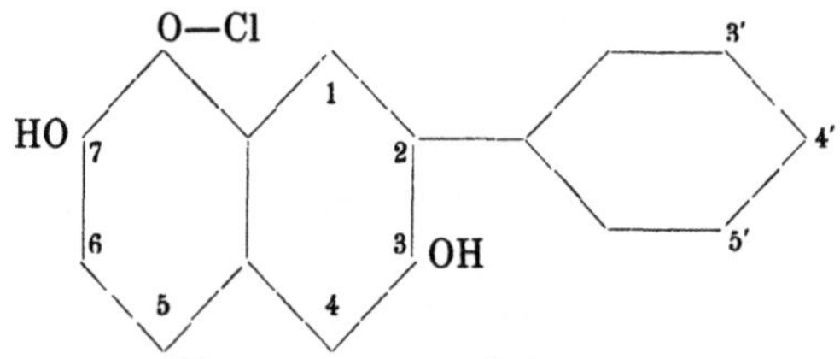

Aglucone

act as auxochromes, the colour tending from red to blue with their increase. Methylation, to OCH_3 at 3′, 5′ and 7 tends to influence the colour by blocking the OH and putting it out of action, thus tending towards red. The glycosidal residue, at 3 and/or 5, if monoglucose $—O—C_6H_{11}O_5$ tends towards reddening, but if diglucose $—O—(C_6H_{11}O_5)_2$ it tends towards blueing. Peonidin is like cyanin methylated at 3′; while delphinidin methylated at 3′ gives petunidin, at 3′ and 5′ malvidin, and at 3′, 5′ and 7 hirsutidin (see SCOTT-MONCRIEFF, 1936).

BLANK (1947, p. 249) gives a Table of properties of six anthocyanidins which includes the following points, using the colour abbreviations as for the above Table IV (Tables I–III):—

	Pelargonidin	Cyanidin	Delphinidin
Colour in aqueous soln.	R	vR	BR
in soda soln.	B	v → B	v → B
Fading in aqueous soln.	on standing	on heating (isomerisation)	slow-cold rapid-hot (isomerisation)
	Peonidin	Malvidin	Hirsutidin
Colour in aqueous soln.	vR	vR	vR
in soda soln.	v → B	v → gB	v → gB
Fading in aqueous soln.	on heating	on heating diluted soln.	on heating diluted soln.

The glycosidal anthocyanin derivatives of these anthocyanidins would tend to be redder with monoglycosidal side groups and to be bluer with diglycosidal side groups, so that one has, theoretically, a considerable number of possibilities, in the absence of direct analytical evidence.

The colour data in BUXTON and DARBISHIRE's Table I tend to support their own suggestion that the yellow colour component giving the greens above pH 8 is a flavone rather than any anthocyanidin derivative. There is only one gB at pH 11, for *Iris Kaempferi* which is red only at pH 3, and blue over a wide pH range which is likely to cover the pH range of aqueous violet-red solutions of malvidin and hirsutidin which go to greenish blue in alkali. The blues at pH 7 and the reds and pinks at pH 3 are very regular, so that different concentrations of the same or very similar anthocyanins could account for the whole of the slight variation shown within Table IV. Further, if aqueous solutions of (? water-insoluble) anthocyanidins are around pH 7, then the only one of the six which can be concerned is delphinidin because the other five are red or violet red, according to this Table from BLANK's data which does not check up very well with SCOTT-MONCRIEFF's colours with soda, see later. With diglycosidal anthocyanin derivatives the violet red of four of these anthocyanidins might become purple, but it is unlikely that they would come near the pure blue at pH 7 given on BUXTON's colour-plate as typical for these "blue flowers."

On a similar comparative basis, the colour data in BUXTON and DARBISHIRE's Table II indicate the occurrence of derivatives of the red pelargonidin, vermilion at pH 3 in all the cases included in this Table II, with the exception of *Dahlia* which shows no trace of the alkaline blue element of pelargonidin derivatives, as suggested here, but pelargonidin anthocyanins are said by SCOTT-MONCRIEFF not to show the pure blue in alkali which is suggested by BLANK's unqualified 'blue' for pelargonidin. BLANK's colours are given as for the aglycones or anthocyanidins.

BUXTON and DARBISHIRE's Table III has such varied types of colouring that any or all of the main anthocyanins may be included within the two sub-groups A and B; but sub-group C (acid) seems to include only one or at most two anthocyanidin derivatives. The group of white flowers examined seem to be of two kinds, *Iris kaempferi* differing from the other three which are all very similar in the pH range of their 'white' and also in the grading of pale yellows (py) to deep yellow (dy) which is not reached by the odd case even at pH 11. Tables I–III as repeated here (Table IV), as well as SMITH's 1933 colour data, were discussed later by SCOTT-MONCRIEFF and her conclusions are tabulated below in Table V.

SMITH (1931) returned to flower colours as natural indicators, comparing colours in the living cells with the colours given by extracting the pigments direct into buffer solutions at various pH values. The extractions were done by putting only the expanded part of corollas into the buffer solution and boiling for ten seconds: on cooling, the tissue was practically colourless, and the coloured liquid was decanted into a fresh tube and sealed. Other more elaborate extraction methods were compared but no difference could be detected in the results. The colour comparisons were made under a microscope so that the cells could be examined individually. The effects of light and temperature on flower colours showing diurnal changes were also investigated: with *Ipomaea learii* higher temperature increased the rate of colour change while the buds were deeper red with opening in darkness: the development of the full blue is conditioned by both light and temperature. Reversible colour changes, from red-more-acid to blue-less-acid and back, were obtained with carbon dioxide in water, as in earlier experiments. The natural pH values for corolla cell-sap of three forms of *Primula sinensis* were determined as about pH 3.1 for Etna, about pH 4.0 for Reading Ruby, and about pH 4.7 for Czar; and at the same pH values in the same buffers the extracted pigments from all three forms gave colours which were not easily distinguishable one from another. SMITH (1933) distinguished three different pigmentations in these three primroses but they were all "magenta" at pH 3.2 and showed 2 or 3 different shades at all other pH values used. SMITH favours the presence of two pigments, perhaps present in 3 different proportions.

A similar appearance of pH regulation is indicated for Mendelian factors governing the epistatic series—crimson, scarlet, port, claret, mauve—in the various colour-strains of *Papaver rhoeas.*

In a subsequent exposition of the technique of calibration of flower colour indicators, SMITH (1933) again used the system of direct extraction of the pigment into buffer solutions, reducing the period of boiling to five seconds and using pressure with a glass rod to assist extraction. The extraction of flavones was reduced by omitting parts of petals where flavone content was known to be high, as in the region round the 'eye' of primrose corollas and the black spot near the base of poppy petals. The time for cooling, after the brief immediate boiling, was standardised to twenty minutes, before the coloured fluid was decanted and sealed in a small

Pyrex tube. Fading was minimised by this procedure, but it still occurred in the range pH 8 to 11. The short boiling did not affect the pH of the buffer solutions used, which were mainly B. D. H. Universal Buffer mixture with SÖRENSEN's phosphates and PALITZSCH's Borax-Boric-acid (1916) used to check that the type of buffer used did not make any difference to the colour obtained with the pigments over the pH range involved. Other methods of getting the pigments into solution gave less satisfactory results. PANTIN's method (1923), which is described, was used for obtaining an accurate colour match between the coloured cells of the petals and the tubes of standard extracted colours viewed side by side by transmitted light.

In her 1933 paper SMITH gives revised pH values for the cell-sap of the three strains of *Primula sinensis*—Etna 3.1 (3.1), Reading Ruby 3.4 (about 4.0), Czar 5.2 (about 4.7). The natural colours and pH of petals are tabulated for 10 reds, 9 magentas, and 8 blues. Three pure blues occur in the last group—*Delphinium tatsiense* pH 7.3, *Cineraria* pH 7.4, and *Ipomaea learii* pH 7.8: all the other pH values are within the pH range 3.1–6.4. As in BUXTON and DARBISHIRE's Tables the pigments were grouped as:—"blues" which were blues of various tints at pH 6.2–7.4, but reds of various tints at pH 1.2–3.1; "reds" which were various shades of red at pH1.2–3.2 and passed to various violet or brown shades at pH 9.6 but never showed pure blue; "magentas" (intermediates) again various 'reds' at pH 1.2–3.1 but purple or violet at pH 6.2 or pH 7.4. Because of the varied terminology of tints SMITH's Tables does not lend itself to contraction, but this variation indicates the complexity of flower-pigmentation, and the tabulated data are records of the very distinct variation of the colours given by the pigment complexes at different pH values: there is only one case in which the same description is applied over more than two of the pH values—1.2, 3.2, 4.4, 5.2, 6.2, 7.4. It follows that pH is a major determining factor in the natural colouration of flowers, although many other factors undoubtedly play their parts. The natural colour *alone* is no certain indication of the pH of the cell-sap, and with a given pigment complex the virage produced may or may not change with differences of up to two pH units, but with a known pigment complex which has been calibrated over a range of pH values and has shown sensitivity to pH changes, the natural pH of the cell-sap and its changes can be estimated with some of accuracy. The various opinions about the *relative* importance of pH as a factor in flower colouration evidently arise from a difference of emphasis along a chain of processes where different concentrations of *interest* affect *opinions* on the relative interest or importance of the many factors acting on different links of what is after all only one chain.

This point of view is well illustrated by SCOTT-MONCRIEFF (1936) who was not impressed by pH as a factor in flower colouration as known in 1929, but in 1936 wrote "However, two independent proofs of the occurrence of pH differences have now been established." She also states that she has made a special study of "the much discussed pH factor," and gives her own results with a wealth of biochemical knowledge about the anthocyanins, combined with their genetical behaviour, which brings some kind of order out of the chaotic fragments of information available up to the date of her 1936 exposition.

Since all anthocyans are some kind of red (oxonium salts) in very acid solutions (pH 1.2–3.2) SCOTT-MONCRIEFF (1936) points out that the classification into 'Reds', 'Blues', and 'intermediates' or 'Magentas', used first by BUXTON and DARBISHIRE (1929 b) and continued by SMITH (1933), is really an arrangement of flower pigments by their colours in neutral or alkaline solutions, that is by the colours of their sodium or potassium salts. These colours are dependent upon the kind of anthocyanidin or aglycone and also upon the positions and numbers of sugar residues and upon methylation, with colloidal phenomena and co-pigmentation with tannins and flavones acting in some cases. The general position with relation to pH colour changes can now be put briefly. All anthocyanins are reddish, in the form of oxonium salts, at pH values of 3 or less: some anthocyanins are pure blue at pH 7–8, others are red, and there are many variations between blue and red; with greens appearing when yellow flavones are present in alkaline solutions. Flavones and flavonols, as anthoxanthins, are rather similar to anthocyanins, the only fundamental difference (*op. cit.* p. 122) being the substitution of oxygen for the (assumed) hydrogen at position 4 in the aglycone diagram given earlier in this chapter: the ivory

Table V.

3-g = glycosides (monosides, biosides, or saccharides at position 3).
3-5-g = diglycosides with sugars at positions 3 and 5.
R = red. B = blue. p = pure. y = royal. V = violet.

Anthocyanins	Examples	Colours with	sat. soln. Na_2CO_3	soln. NaOH	solid NaAcetate
1. Pelargonidin 3-g	(callistephin)		VR	VR	VR
2. Peonidin 3-g			V	V	VR
3. Cyanidin 3-g	(chrysanthemin) (antirrhinin) (mekocyanin)		BV	pB	VR
4. Pelargonidin 3-5-g	(pelargonin)		RV	V	BR
5 a. Malvidin 3-g	(primulin)		VB	VB	V
5 b. Malvidin 3-g	(saccharide)		V—VB	—	—
6. ? Petunidin 3-g	(saccharide)		V—VB	—	—
7. ? Delphinidin 3-g	(saccharide)		V—VB	—	—
8. Peonidin 3-5-g	(peonin)		(p)B	pB	VR
9. Petunidin 3-5-g	(petunin)		pB	pB	BV
10. Delphinidin 3-5-g	(delphin)		pB	pB	VB
11. Cyanidin 3-5-g	(cyanin)		pB	pB	V
12. Malvidin 3-5-g	(malvin)		pB	pB	V

1-2-3 = "Reds." 8-9-10-11-12 = "Blues."
4-5-6-7, 9 and possibly 8 = "Magentas" or "Intermediates."

anthoxanthins act as co-pigments for anthocyanins, while the yellow-flavones give their own yellow as a component of green at pH values above 7.

Some of the data in SCOTT-MONCRIEFF's exposition can be tabulated so that they connect up with SMITH's colours (1933) and with those of BUXTON and DARBISHIRE which are given above in Table IV, by re-arranging her Table I (*op. cit.*, p. 121) in relation to her suggested grouping of the 'reds', 'blues', and 'magentas' (*op. cit.*, p. 132).

Numbers 5 b, 6 and 7 do not occur in SCOTT-MONCRIEFF's Table I (*op. cit.* p. 121) and it seems advisable to quote her own text on this matter (*op. cit.* p. 132):—"A study of the colour changes with sodium acetate and sodium carbonate" in Table I shows that the 'red' group would comprise the 3-monosides and 3-biosides of pelargonidin, peonidin, and cyanidin, whose reactions range from violet-red to violet, while in the blue group can be placed the 3-5-diglycosides of cyanidin, peonidin, delphinidin, petunidin, and malvidin, which all give a pure blue with sodium carbonate. In the 'intermediate' or 'magenta' class are pelargonidin and petunidin 3-5-diglycosides, also malvidin 3-saccharides and probably petunidin and delphinidin 3-saccharides, whose carbonate reactions range from violet to violet-blue but are near pure blue. Recent chemical analysis of the pigmentation of the flowers studied by these authors is in close agreement with this explanation. Modification by co-pigments must also be taken into consideration."

Reference to Table V above shows that VR (violet-red) with solid sodium acetate characterises all the "reds" (1–3) and that pB (pure blue) with sodium hydroxide does the same for all the "blues" (8–12). But cyanidin 3-glycosides are given in SCOTT-MONCRIEFF's Table I (*op. cit.* p. 121) as "pure blue" with NaOH and as "violet red" with solid sodium acetate and "the cleared extract" used: the peonidin 3-5-glycosides also give pB with NaOH and *vivid* violet-red with sodium acetate, so that there is some colour overlap for the cyanidin 3-glycosides in the red group and the peonidin 3-5-glycosides in the blue group: the latter are given in Table I as only "blue" with Na_2CO_3 and as "pure blue" with sodium carbonate in the above quotation. The violet to violet-blue and reddish-violet (RV) in the "magenta" or "intermediate" group, with NaOH or Na_2CO_3, are not clearly differentiated from those in the red group by these present colour-data, although the latter seem to be separated by the violet-reds with sodium acetate. The blue-red of pelargonidin 3-5-glycosides with sodium acetate seems very appropriate to an "intermediate" group, and so does the reddish-violet with Na_2CO_3.

These colours with Na_2CO_3 and NaOH are not likely to correspond to natural colours in the cell-sap of flowers, since nearly all juices expressed from the coloured parts of petals have been described as "acid": but they do give a classification based upon the known pigments in the three colour classes. These pigments under natural conditions may be subject to the complications of mixture of pigments, of colloidal conditions or the presence of co-pigments, and so the natural virage may not be that shown by the pure pigment. The greens, shown at pH 11 by nearly all the cases in BUXTON and DARBISHIRE "intermediates," A, B, and C-groups, indicate the presence of flavones which can act as co-pigments with anthocyanins and thus modify the colours shown at lower pH values.

Thus agreement seems to have been reached in 1936 that pH does play its part as a recognised factor in flower colouration. SCOTT-MONCRIEFF (1936) even gives cases where the pH value of the coloured cells of petals is determined in the first instance by a Mendelian factor or gene, as for example in the strictly localised acidity in the petals of *Primula acaulis* with its factor **S**:— "the author is in agreement with Mr. BUXTON's suggestion that **S** governs the pH of the cell sap" (*op. cit.*, p. 148). Also, *loc. cit.* p. 130—"In *Primula sinensis* all cyanic flowers show a pH differentation according to whether they carry **R** (pH 5.4) or **r** (pH 6.0)" and "In *Papaver rhoeas* a similar differentation between the **P** genotypes (pink, scarlet and crimson) with a pH of about 4.9 and the recessives (mauve, petunia and port) with a pH of about 5.8 is also unconcerned with the anthocyanidin type. The pigment of the dominant acid form, however, is complex, some organic acid being combined with the pigment molecule." In some other cases, such as *Aquilegia* no such pH differences were found, but "only dried material was tested."

In spite of SMITH's work on a method which does approximate to an estimation of natural pH values within living cells of coloured petals, other workers were satisfied with the pH values obtained with a glass electrode applied to the mixed juices expressed from the coloured parts of corollas or petals. These pH values need not be the same as the natural values but they seem to follow sequences and distributions similar to those of the natural values and so yield instructive *correlations* between pH values and some flower colours. Cases which have been described in which two different colours are shown at the same pH are naturally to be explained as due to factors other than pH differences, but such explanations and factors are not relevant in this chapter on "pH and natural indicators."

Since SCOTT-MONCRIEFF's 1936 exposition of this subject the work on pH in relation to anthocyanins has been confined to particular cases and BLANK's extensive list of 594 references (1947) includes very few later than 1939, with no title that suggests any change in the general position as outlined above.

IV. pH of Plant Cell-Sap

(a) Early Methods and Results

Published records of pH in plants were, before 1926, mainly concerned with the estimations of pH for fluids or juices which had been pressed out from various parts of plants and examined by means of indicators or by electrometric methods. As this contribution to *Protoplasmatologia* is more or less restricted to pH in relation to the contents of living cells as they occur *inside* the cells, the present significance of the pH values of expressed mixed juices is rather difficult to assess. Readers desiring to consult early literature about pH in plant juices are referred to the bibliographies and some of the 'special applications' given in the 1920 and 1927 editions of CLARK's classic on *"The Determination of Hydrogen Ions,"* and to the early history of pH of plant cells and tissues given by SMALL (1929). ROHDE (1917) and ATKINS (1922) were among the first to use synthetic indicators on immersed plants and sections of plants. As one very clear example of the use of *tints*, ATKINS' 1922 description of his results is instructive thus—"Using methyl red, the sclerenchyma and bast fibres give a salmon pink to pink colour, corresponding to pH 5.4 to 5.2; the wood appears a faint pink, not more acid than pH 5.4 to 5.6 and the medullary rays are yellow. The upper limits of acidity are therefore fixed. It remains to determine the acidity of the portions appearing uniformly yellow of the same tint with di-ethyl red. With phenol red the full yellow appeared, showing an acidity of pH 6.6 or more. With bromthymol blue a yellow with a green tinge was seen, matching the colour with pH 6.2". In order to avoid an error here due to a green tinge in the tissue a drop method with expressed sap was used. This tint method applied to fresh sections of *Salvia verbenacea* gave pH 5.2–5.4 for the sclerenchyma and bast fibres, pH 5.4–5.6 for the wood walls and pH 6.0 for medullary rays, medullary and cortical parenchyma.

When one considers—1. that a carefully prepared standard indicator colour is generally regarded as necessary for accuracy even to .2 in pH values: 2. that the concentration of the indicator in the solution to be tested must not differ materially from that in the standard solution: and 3. that both solutions are usually placed in tubes of standard size and viewed in a comparator which permits a view only through the whole thickness of the fluid, one is rather surprised at the accuracy which is assumed by the publication of results, old and more recent, of this method of using indicators on sections. The possible errors involved are numerous and include—1. variation in depth of colour due to variation in the thickness of the sections; 2. variation in depth and/or quality of tint from tissue to tissue due to variation in the rate of penetration of the colouring fluid; 3. variation in the quality of tint due to sap which has escaped from cut cells; 4. the 'protein' error; and 5. the 'salt' error.

The first two sources of error are so obvious to a critical investigator that CLARK (1921, p. 118) referring to "the use of unequal depths of solutions through which the colours are viewed" wrote that "There are errors of technique ... which may be passed over with only a word of reminder." He also refers to certain optical effects (*ibid.* p. 65), *e. g.* where a dichromatic purple such as that of bromophenol blue may appear red or blue according to the thickness or thinness of the layer of fluid traversed by the light examined. This optical effect might explain some of the records of blue-red colours mentioned above in connection with flower colouration.

With regard to the second source of error we have in fact found tissues which, although giving a stronger red with ethyl red than surrounding tissues, give a yellow with methyl red while these surrounding tissues give a red with methyl red, and similar results have been obtained in other cases with other indicators. The depth of the tint observed is, therefore, no certain guide to the pH value for a tissue or single cell.

The third source of error can be avoided by suitable washing with neutral unbuffered water. The fourth or 'protein' error has its source in the removal, by adsorption or otherwise, of the indicator from the field of action and must be considered in the case of the living cell. The fifth or 'salt' error may amount to $+0.35$ with bromophenol blue and .5 N sodium chloride but is not likely to be greater than ± 0.1 for plant cell-sap, and is likely to be less than 0.1 with all the indicators used by ATKINS.

In spite of the possible errors of tint-methods used on fresh sections, the results of such methods of estimating pH values of plant cells can be re-interpreted in terms of the R. I. M. interpretation of colours as distinct from tints. The sectioning technique is surely more reasonable and more likely to give a closer approximation to the truth concerning significant variations in pH within the plant than are the electrometric methods which have been used. The nearest approach to an unmixed sap used for such determinations is that of *Valonia*, used by CROZIER (1919), BROOKS (1926, 1930), and others. But considering the comparatively extensive variation which has been demonstrated in the pH of neighbouring cells, the determination of the pH of a liquid which is a crude mixture of cell wall fragments, cytoplasm and vacuolar sap with most of their inclusions, does not seem capable of providing very satisfactory data upon which to base conclusions concerning the role of pH values in the life of the plant. The data obtained by CROZIER (1919) on the pH of liberated vascular sap might be more reliable and more significant than electrometric pH values for mixed juice, but even in that case the very process of liberation involved some change in the pH of the fluid, as was shown later by BROOKS who in 1930 revised her own 1926 data on pH and rH of *Valonia* cell-sap after reconsidering the carbon dioxide factor. In the case of a very slightly buffered cell-sap it has been demonstrated that the carbon dioxide content can be a significant factor in controlling the pH of the sap, and wounding the cell may profoundly alter the carbon dioxide content (LILLIE, 1909): also allowing the carbon dioxide of the sap to attain an equilibrium with the air will certainly alter that factor (cp. LIMPEROS and RANTA, 1953).

The methods, still used occasionally, which involve the expression of the sap, seem to the writer to have somewhat the same significance as would determinations of the pH of an average chemical laboratory by a process which involved as its first step the crushing and mixing of all the receptacles and their contents. Quite apart from the inevitable interactions, the resulting data, while more or less true as a kind of average, would furnish no very obvious clues as to the processes included in the normal activities of the laboratory. Nevertheless, those who use these methods work out their data to the second decimal place, and some have even claimed a significance for variations in the last figure of such data. If, however, the expressed sap is from a homogeneous tissue, useful information can be derived from such data: and if the same kind of organ such as a leaf is used, even with its differentiated tissues, *comparative* values can be obtained, which again may be of use for comparisons within a given series of observations, such as in a series of allied species or in a series of observations on the same species under different natural or experimental conditions.

(b) R. I. M.

SMALL's Range Indicator Method (R. I. M.) has been described in print several times (SMALL 1926, 1929, 1946, 1953 a) and has recently been written up in considerable detail with all precautions and the reasons for these in SMALL's contribution to SPRINGER's (Berlin) handbook on *Modern Methods of Plant Analysis,* and is repeated here, with their permission, as being essential for a proper appreciation of what follows. The method has been described as "laborious," but "Truth is worth Labour," even if the truth is trapped only within specified limits.

Results. During the first five years, when the R. I. M. was being tested without any publication, two colour conscious workers, Miss M. W. REA and Miss S. H. MARTIN, made various surveys including the following:— 1. a preliminary survey of the general pH of tissues in flowering and other stems; 2. a complete survey of the tissues of the sunflower *(Helianthus annuus)* from the dry seed to the mature flowering stage; 3. a similar survey of the broad bean *(Vicia faba)*; 4. a survey of tissue pH values in the stems and leaves of herbaceous and woody plants for each month in the year; 6. a survey of tissue pH values of selected species in markedly different habitats; 7. a survey of pH values in succulents; 8. a more extended survey of pH values in flowering stems: also 9. investigations by MARTIN of the buffer systems in sunflower and broad bean. Later work included a survey of pH values in the potato by SMALL to supplement investigations of the buffer complex in potato and other plants by INGOLD; and a survey of pH values and buffer complexes in fungi by ARMSTRONG. All these were published in detail in *Protoplasma* and, again in considerable detail, in *Protoplasma Monographs,* No. 2. Here they are summarised with the new notation under appropriate aspects of "pH of plant cells."

(c) Plant Juices

The pH values recorded for expressed juices of plants are *average* values and depend upon the original natural values in differentiated tissues and cells. The average will also depend on the relative development of more acid and less cell-sap, and on the methods of expression and manipulation of the juice before and during the estimation of the pH values. The recorded values are therefore somewhat aleatory, but they may have some significance, especially in the case of very low pH values, and in cases where the pH of expressed juices vary with external conditions.

Heyne (1815) used the criterion of sour taste on the diurnal variation of pH *in some tissues* of the leaves of *Bryophyllum calycinum*, a phenomenon which has very modern aspects. Ulehla (1928) noted a variation in the pH of the pressed sap of *Opuntia phaeacantha* within the range of pH *in some tissues* of the leaves of *Bryophyllum calycinum*, a phenovalues, some cells in this *Opuntia* must have a sap content of higher acidity than pH 1.4, so that the secreting living cell is capable of action at very high acidities, in plants as in animals. The pH values of gastric juice in adult humans are about 1.6–1.8, with hyperacidity at 1.4.

One of the most striking cases of pH values depending on manipulation and procedure during the estimation of pH of a plant fluid is Hempel's (1917) study of the pitcher fluid of *Nepenthes.* She found a variation of pH during one determination by a hydrogen-electrode method from pH 5.87 to pH 8.15, and attributed the rise in pH to an effect of the hydrogen gas bubbling through the fluid.

Gaudichaud (1848), Sachs (1862) and Kraus (1883–1886) were early observers of the normal acidity of plant saps: they used litmus. Wagner (1916), A. R. C. Haas (1916–1920), Hempel (1917), Rohde (1917), Kappen (1918–1919), Truog (1919) and Clevenger (1919), were among those who investigated pH of plant juices at an early stage in the history of this subject. Their results were used by Small (1929) in his survey of the early literature and some of them are still applicable to some aspects of pH in plant cells. But the present contribution is more or less restricted to the pH of actual contents of living cells in accordance with the general title of Protoplasmatologia.

(d) pH of Cell-Sap within Cells

1. Systematic Data

Algae. Rohde (1917) used very diluted indicators, 0.5 per cent stock solutions diluted (1 in 500); the indicators used were neutral red, methyl red, and methyl orange, and a plasmolysis test was applied to check the final living condition of the cells in immersed fresh whole material or sections. He also made observations on pressed juices with a H-electrode. *Spirogyra* cell-sap is reported as pH 5.5–5.9 during the day and pH 6.7 during the night. Lapicque (1921–1923) found an interior pH below 5 in *Spirogyra* in spite of a surrounding medium at pH 8. Pearsall and Ewing (1925) give pH 6.2–7.0 as the normal sap reaction of *Spirogyra.* The later authors do not record the time of day of their determinations and Rohde's records seem to be more critical, although he used the *tints* given by neutral red as a guide and reported that the H-electrode yielded similar results. This acidification of cell-sap during the day is unlike later observations with higher plants, and Rohde *(op. cit.)* reported similar day-acidification in the epidermal cells of carnation and tulip flowers for which he interpreted indicator tints as showing pH 5.9 down to pH 4.9. Rohde *(op. cit.)* used a third method for estimating the pH status of cells, namely the speed and degree of absorption of basic and acidic dyes. With *Spirogyra* basic dyes gave distinct colours in a short time (a few seconds up to a minute), while acidic dyes gave no colour. He compared this with more acid cells of unripe apples and of rhubarb leaves which with basic dyes colour slowly but distinctly while with acidic dyes they colour very quickly and intensely.

Atkins (1922) using his section method with tints of various indicators records pH 7.2 for the medullary tissue of the receptacle of *Fucus*; pH 6.6 for the stipe and pH 6.9 for the disc of *Himanthalia*; pH 7.0 for *Ulva* and for the diatom *Skeletonema*; and pH 7.3 for *Laminaria* stipe. These relatively high pH values are not in very close agreement with those found by Clapham (in Small, 1929, p. 93–94), who used Small's R. I. M. and reported the following pH values for the cell-contents and walls of the ventral region of the thallus:— *Laurencia caespitosa* pH 4.0 *ca.*; *Laurencia pinnatifida* pH 4.0 *ca.*; *Chondrus crispus* within the range 5.9–4.4; *Dictyota dichotoma* within the range pH 6.2–4.4. Some of the indicators were not absorbed by these algae, but the *Laminaria* material did show some differentiation as set out below.

Laminaria digitata. R. I. M. pH ranges.

52–48 young and old stipe—outer cortex: medium aged stipe—both outer cortex and medulla.
59–56 medulla of old stipe.
59–59 outer cortex and medulla of some young stipes, and outer cortex of young and old hapteron.
62–59 medulla of some young stipes.
62–62 inner cortex of some young stipes.
68 62 medulla of old hapteron and inner cortex of old stipe.
68–64 inner cortex of some young stipes and of medium-aged stipes which also showed a band at 52–48, also medulla of young hapteron.

Laminaria saccharina. R. I. M. pH ranges.

52–48 outer cortex of very young stipe.
56–56 medulla of very young stipe.
59–59 inner cortex of very young stipe, outer cortex and medulla of young stipe.
68–64 inner cortex of young stipe.

The differentiation here is distinct. The outer cortex of both species varies from acid (52–48) to less acid (59–59). The inner cortex is below pH 6.0 only in the very young stipe of *L. saccharina,* otherwise it varies from 62–62 in some cases to a more general 68–64 or almost neutral. The medulla is more variable than the cortex, being least acid in the hapteron (68–64 young, and 68–62 old), and 62–59 or 59–59 in the young stipe of *L. digitata,* 59–56 in the old stipe of that species, 56–56 in the very young stipe of *L. saccharina* with 59–59 in the older (young) stipe of that species; the most acid value recorded for the medulla was 52–48 for the medium-aged stipe of *L. digitata.*

There is considerable differentiation in these R. I. M. pH ranges, total range 52–48 to 68–64, but no value above pH 6.8. Atkin's record of pH 7.3 for *Laminaria* could be correct, keeping in mind the variation from 5.9 to 8.1 obtained by HEMPEL (1917) during one electrometric estimation of one fluid. The whole range, pH 5.2 to pH 7.3, is within the influence of carbon dioxide on the pH of cell-sap.

Fungi. There are very numerous records of the pH values of culture media and the effects of pH on the germination of fungal spores and the growth of fungi (see SMALL, 1946, 1953), but there are few records of the intracellular reaction of fungal hyphae. HERRMANN (1879) noted that the plasmodium of *Aethalium* turned red litmus paper blue, but that may not mean a pH value above 7.0 (it depends on the type of litmus) although real alkalinity is quite probable. INGOLD (see SMALL, 1929, p. 94) using a drop comparison method and applying the R. I. M. method of interpretation to the colours obtained, estimated the pH of the cell sap exuded on puncturing the sporangiophore of *Pilobolus crystallinus.* This sap was alkaline to the whole range of R. I. M. indicators, from BPB to PR (>7.0), but yellow with cresol red (<7.4). The actual pH lies, therefore, in the range pH 7.0–7.4.

ARMSTRONG (1929 a) made a good beginning towards removing this gap in our knowledge of intracellular pH in fungi, but there is still a large gap. He also investigated very successfully the buffer-complexes of some selected species. The pH values were usually in the pH range 62–56 with little differentiation in the fructifications of both Basidiomycetes and Ascomycetes.

R. I. M. pH *ranges for Fungi* (after ARMSTRONG, 1929 a).

e 52–48 occurred in the stipe of *Hypholoma fasciculare.*

c 56–56 was recorded for the hymenium in young fructification of *Armillaria mellea,* for the pileus and hymenium of *Lactarius blennius,* and for the hymenium and some hyphae of the stipe in the ascomycete *Leotia chlorocephala.*

b 59–56 occurred in stipe, pileus, and hymenium of *Coprinus micaceus* both young and mature, and in the pileus and hymenium of *Panus torulosis.*

a 59–59 was commoner, recorded for stipe, pileus, and hymenium of *Mycena vulgare, Mycena pura, Psalliota campestris, Polystictus versicolor, Polyporus* sp., mature *Armillaria mellea* and for the stipe and pileus of young *A. mellea;* for the pileus and hymenium of *Hypholoma fasciculare, Clitocybe laccata, Collybia radicata,* and for the stipe in *Typhula incarnata* and for the hymenium of *Cortinarius violaceus* and the conidial region of the ascomycete *Xylaria hypoxylon.*

C 62–62 was given for the stipe, pileus, and hymenium of *Amanita muscaria* and some specimens of *Armillaria mellea*; for stipe and hymenium of *Clavaria rugosa, Clavaria corniculatus,* and the ascomycete *Helvella crispa;* for stipe and pileus of *Cortinarius violaceus;* for the stipe of *Clitocybe laccata, Panus torulosis,* and some hyphae in the stipe of *Leotia chlorocephala.*

D 68–62 occurred in stipe, pileus, and hymenium of *Coprinus atramentarius*..

Apart from the acid (52–48) stipe of *Hypholoma fasciculare* and the nearly neutral whole fructification of *Coprinus atramentarius*, all the other fungal cells examined had pH values in the range 62–56.

Armstrong (1929 a) also investigated the buffer index values of all the larger basidiomycetes in the above list and found that the maximum buffer index values for *Coprinus atramentarius* and *Polyporus sulphureus* occurred between pH 5 and pH 6, and both were very low, .004 for the former and .006 for the latter. The maximum β (buffer index) values for eight others occurred between pH 4 and pH 5, and some of these were much higher: in order of increasing buffer capacity as measured by maximum β values these were:— *Polystictus versicolor* .001, *Lactarius blennius* .006, *Lycoperdon pyriforme* .007, *Hypholoma fasciculare* .009, *Coprinus micaceus* .010, *Collybia velutipes* .026, *Armillaria mellea* .027.

The buffer complex of *Coprinus micaceus* was found by analysis to consist mainly of 0.025 M oxalate, with 0.006 M phosphate and 0.0026 M malate. Using the same system of analysis, the buffer complex of *Collybia velutipes* was more or less completely analysed as 0.0228 M malate, 0.014 M citrate, and 0.010 M phosphate, which is qualitatively very similar to other buffer complexes such as are known in higher plants. The quantities of buffering substances found by analysis in the expressed sap of these two species accounted for most of the buffering from pH 4.0 to pH 7.0; 75–100 per cent in *Coprinus* and 85–100 per cent in *Collybia*. Van Slyke's buffer index (1922) for titration of weak acids with a strong base is dB/dpH where dB is the amount of strong base added for an observed shift in pH = (dpH). Buffer index curves are discussed mathematically by Ricci (1952). These buffer index curves can be calculated from detailed titration data, and used to check the buffering capacity of cell-saps from *homogeneous* tissues with some relevance to cell contents, see Small (1929, 1953) on analysis of separated tissues with complementary distribution of malate and phosphate by Thoday *et al.* (1931–1939). Notwithstanding all the approximations used in measuring pH values during titrations and in the processes used for analysis, the combined β values of malate, citrate, and phosphate between pH 4 and pH 5 accounted for 97 to 101 per cent of the maximum buffering capacity of *Collybia* sap in the pH range, 4.0–5.0. A similarly close analysis of *Pelargonium* sap by Armstrong (1929 b) yielded a buffer complex consisting of 0.0537 M oxalate, 0.0337 M malate, 0.0066 M tartrate, 0.010 M phosphate, and a metallic hydroxide, which account quantitatively for from 81 to 101 per cent of the buffer capacity of fresh expressed juice of young stems of *Pelargonium* in zone of maximum buffering, pH 3.4 to pH 4.4 (*op. cit.*, p. 337).

The degree of buffering present in plant cells around the natural pH of the cell sap has a considerable influence on the strength of carbon dioxide effects between pH 5 and pH 7. These aspects of the control of pH in plant cells are discussed later.

Bryophyta. Although a large amount of work has been done on the distribution of bryophytes in relation to the pH of their immediate habitats, there are few records of intracellular pH values for this group. There were none in 1922, and Dr. M. J. LYNN (see SMALL, 1929, p. 95–96) investigated the possibilities. Differentiation in the pH of cells was difficult to find, but in *Marchantia polymorpha* and *Fegatella conica* both types of rhizoids were at pH 40–40, while the general thallus was at 44–44 in the latter and less acid (52–48) in the former. More differentiation was found in *Polytrichum commune*: 44–40 was noted for the cell contents of the epidermis and cortex of the leafy stem; the hydrome mantle cell contents varied from 44–40 to 52–44 and the leptome was 62–59, while the cell-walls of both epidermis and hydrome showed a very acid virage ($<$ 3.4).

Pteridophyta. There were no records in 1922 of intracellular pH values in this group, and again Dr. M. J. LYNN sampled a few cases (see SMALL, 1929, p. 96). In *Equisetum maximum* there was self-colour in the epidermis, outer cortex, endodermis, and pericycle of the stem internode, but pH ranges were observed in other tissues:—

40–40 was found in rows of cells in the sporangiophore stalks and some cells of the peltate head:

52–48 was common, occurring in the cells of the pith and in the cells lining the vallecular canals in rhizome, stem, and strobilus, also in pericycle and endodermis of rhizome and strobilus, and in the outer cortex of the strobilus axis, where there was not self-colour as in the aerial stem and rhizome:

56–48 occurred in the inner cortex of the rhizome: and

56–56 in the same part of the young strobilus axis:

59–56 in the phloem of the stem internode:

59–59 in the phloem of the rhizome and strobilus axis: with

62–62 in the protophloem of all parts of the plant.

The various tissues of *Equisetum maximum* cover the same pH range, 62–40, which is normal for flowering plants, with differentiation from tissue to tissue.

Selaginella martensii is more limited: 52–48 occurs throughout the cortex and epidermis, and 59–59 in the pericycle and phloem, with a very acid virage (40–40) in the walls of the xylem.

Dryopteris filix-mas is even less varied with the range 52–48 throughout all the rhizome and young leaf rachis, except for acidic wall virages in xylem and sclerenchyma.

Asplenium viviparum showed more differentiation both in cross-sections and as maturing changes:— the cell contents of endodermis, pericycle and phloem were 59–59 at the tip of the young rachis, 56–56 at its base, and 52–40 at the base of the old rachis:

56–56 occurred in the cortex of the young rachis: with

52–48 throughout the cortex of the old rachis: but

52–40 in the epidermis and ramenta of the young rachis.

Apart from the last item, acidification of cell contents with age of tissue is indicated for endodermis, pericycle, and phloem (59→56→52–40) and also for cortex (56→52–48).

Coniferales. The only intracellular records here are a few by M. J. LYNN (see SMALL, 1929, p. 97). In *Pinus nigra* f. *austriaca* two years old stem the pith and phloem parenchyma are 52–48, and some cells of the inner cortex 44–44, with the cell-walls of epidermis, hypodermal sclerenchyma, and xylem all showing a very acid virage (< 3.4): in the mainly self-coloured mesophyll of the leaf some cells were clearly 52–48, and the contents of the endodermis < 3.4, like all lignified or cuticularised cell-walls. In *Taxus baccata* f. *fastigiata* two years old stem the cell contents varied; epidermis and phloem parenchyma 52–44, many cortical cells 44–44, and many pith cells < 3.4: in the leaf there were scattered cells with acid contents (40–40) in the epidermis, mesophyll and bundle sheath.

The strong acidity of conifer tissues shown by these R. I. M. data is abundantly confirmed by results obtained by DOYLE and CLINCH (1926), also ROBERTS and DOYLE (1938). DOYLE has published on the conifers from many points of view but the pH aspects are more easily discussed in relation to the 1938 data and methods. ATKIN's drop method was used on expressed leaf saps in 1926, but in the 1938 investigation many methods were tried on these leaf-saps which are difficult to obtain in a clear fluid condition. These methods were checked, one against another, for consistent results. The large proportion of green colour in the mesophyll rendered the R. I. M. impracticable. The method finally adopted was the gas-chain or hydrogen electrode method, the pH results being checked colorimetrically using a B. D. H. Capillator: these two sets of data "almost invariably agreed" within the limits of capillator accuracy. A quinhydrone electrode was used as a second check method in some series of materials. Two methods were used for extraction of sap from leaves, and checks with both methods for the same material did not show more than 0.05 pH as a difference. Method A consisted of grinding 5 g. fresh leaves in a mortar with 20 mils of boiled distilled water and filtering through boiled dried muslin. Some material had too much colloidal matter for this direct process. Method B, following that of DIXON and ATKINS, consisted in freezing the tissues in liquid air, allowing them to unfreeze and then pressing out the juice in a vice between two iron plates which were surfaced with silver foil. This method was used on some materials because it avoided the swelling of colloids which occurred with method A. Both methods involve mixing saps from different tissues, which in LYNN's R. I. M. data vary from 52–48 to < 34 in two species.

Second decimal place figures are used in presenting some of the resulting data on maturing and seasonal changes in pH, which will be discussed later, but in analysing the systematic distribution of pH ranges, an accuracy of ± 0.05 is considered sufficient. A graphical tabulation is given of pH of conifer leaf saps, from which the following distributions can be summarised without the specific names thus:—

Pinaceae: 3.4–4.0 in *Abies* (3 species), *Cedrus* (2 spp.), *Pseudolarix, Tsuga, Keteleeria* (1 sp. each): 3.3–4.35 in *Pinus* and *Picea* (6 spp. each), *Larix* and *Pseudotsuga* (1 sp. each): 5.2 in mature leaves of *Picea breweriana.*

Taxaceae: 4.5–5.8 with *Cephalotaxus* (3 spp.) 4.5–5.1, *Taxus baccata* 5.3, *Torreya* (2 spp.) 5.7–5.8.

Araucariaceae: 3.25–4.4 with *Agathis* (3 spp.) 3.25–3.5, and *Araucaria* (4 spp.) 4.05–4.4.

Podocarpaceae: 3.6—5.7 apparently in two subgroups thus:— 3.6–4.15 *Podocarpus* 2 species of section *Dacrycarpus,* with *Dacrydium* (2 spp.), *Saxegothaea,* and *Phyllocladus* (1 sp. each); 4.5–5.7 *Podocarpus* (11 spp.), and *Microcachrys* (1 sp.).

Cupressaceae: 3.5–5.7 in 5 subgroups which are each of related genera, thus:— 4.5–5.5 *Cupressus* (7 spp.), *Libocedrus* (2 spp.); *Thuja, Thujopsis,* and *Fokienia* (1 sp. each): 4.5–5.7 *Juniperus* (5 spp.); 5.2–5.7, *Widdringtonia* (3 spp.) 4.5–5.4; *Fitzroya, Actinostrobus,* and *Tetraclinis* (1 sp. each): 4.7–5.4 *Athrotaxis* (3 spp.); *Taiwania, Glyptostrobus,* and *Cunninghamia* (1 sp. each): 3.5–4.3 *Callitris* (5 spp.): 3.6–4.0 *Sequoia* (2 spp.), *Taxodium* (2 spp.), *Cryptomeria* (1 sp.). *Sciadopitys verticillata* is ungrouped at 4.5.

The above arrangement by pH ranges *of the grouped genera* is also a taxonomic arrangement. The genus with most species usually covers most of the pH range of the group of genera, but grading and subdivisions can sometimes be seen in the original graphs. *Cephalotaxus* 4.5–5.1 is more acid than *Torreya* 5.7–5.8. *Agathis* is more acid than *Araucaria. Juniperus* is less acid than the associated genera while *Widdringtonia* covers a wider range. There is no doubt in this case that the pH values of mixed saps of leaves are in fact comparative values of considerable interest. The mixed leaf sap of *Taxus baccata* was around pH 5.3, but Lynn found scattered cells with contents at 40–40 in the mesophyll, bundle sheath, and epidermis, so that the green cells of the leaf are probably of higher pH, above 5.3, and the average may result from mixing very acid with nearly neutral saps. The mixed leaf sap of *Pinus nigra* was around pH 3.7, while Lynn reported 52–48 for some mesophyll cells and <34 for the endodermal cell contents so that in this case the acid cell-saps seem to dominate the less acid saps in the mixture.

Angiospermae. pH values for cell-contents had been recorded previous to 1926 mainly by Rohde (1917), Atkins (1922), and Pfeiffer (1925 a) but tint comparisons were used and a re-interpretation of these data based upon the actual colours recorded leaves them incomplete with either a lower or a higher limit lacking for particular cases, except in a few instances with rather wide pH ranges. A general survey was carried out during the five years before 1926 by Small, Rea, Martin, and Ingold, using the R. I. M. Ingold made a special investigation of the effects of various percentages of alcohol (see Ingold and Small, 1928, and Small, 1929, pp. 51–59) as a result of which indicator solutions were used with no alcohol or in any case not more than 10 per cent alcohol. This survey lost its *general* character at an early stage and investigations of particular aspects of the pH of plant cell contents have continued at intervals up to the present time. Atkins (1922) reported pH values as follows:—

Salvia verbenacea—sclerenchyma and bast fibres 5.2–5.4, wood walls 5.4–5.6, medullary rays and parenchyma 6.0.

Taraxacum officinale—stem and leaf-midrib, 4.6, leaf parenchyma 5.8, root stock medulla 5.4, latex canals 4.4–4.8, vascular bundles much the same with tint difference.

Anagallis arvensis—stem vascular bundles < 5.2, pith > 5.2.

Samuel (1927) recorded leaf mesophyll of *Prunus laurocerasus* as 5.6–6.0, usually 5.7–5.8, using a micro-quinhydrone method.

Pfeiffer (1925 a) devised another section-immersion method which eliminated the use of both two-colour indicators and standard buffered comparisons, and later (1927) published a series of pH values for the absciss layer of leaves after frostfall—*Fuchsia* spp. 4.8–5.2, *Begonia* spp. 3.9–4.3, *Azalea* 6.2, *Aesculus* spp. 5.4–5.8,*Echeveria* spp. 6.2–6.8, *Cotyledon* 6.9–7.1. Pfeiffer (1925 b) gave a very interesting account of the form of calcium oxalate crystals in relation to the pH of the medium in which they are formed (see Small, 1929, pp. 99—102 for discussion).

2. R. I. M. General Survey

Rea and Small (1926) recorded data for over 150 kinds of stems, and this series was later extended by these two to 216 stems and leaves of 181 different species (Small, 1929). All tissues which were not self-coloured were examined with the R. I. M., and yielded definite pH ranges in most cases. The full results were published under an alphabetical notation which seemed at that time to be a suitable method of abbreviation, but the intermittent use of this list has shown that a numerical notation with some indication of the pH values concerned is needed for ready comprehension of the significance of the numerous data. Wall virages were included, and these were acidic (52–40, 44–40, or < 34) for all lignified walls as in xylem and sclerenchyma, all maturely cuticularised epidermal walls, and all suberised walls as in cork and endodermis. These are noted here as general data referring to all species in which they occur, but acidic records such as 44–40 can apply to cell-contents both in xylem and in other tissues. The xylem parenchyma may be as high as 59–59 but usually it is 44–40, and that range occurs for procambial strands in the germination stage of the seeds of both broad bean and sunflower, so that acidic lignification of walls may follow an acid cell-sap at 44–40. Cellulose walls seldom show a definite colour with indicators but collenchyma indicates 52–48 frequently, and some other cellulose walls do show definite colours.

In order to get some idea of the distribution of more acidic and less acidic tissues, the records for particular species have been grouped by the pH ranges for more acid and less acid epidermis, with a mixed or varying group between; and these groups have been subdivided in various ways, such as by low pH values in the epidermis (40–40, 44–40, 52–40 in decreasing order of acidity in cortex and pith) to 52–48; and by higher values (56–56, 59–56, 59–59) in the epidermis. This separates some members of the all acid families into different grades of "more acidity," but their occurrence still shows strongly. These 'all acid' families which have most or all of their cell-contents within the 52–40 range are—Polygonaceae: Crassulaceae, Saxifragaceae, Ribesiaceae, and Rosaceae; Geraniaceae, and Tropaeolaceae; Cactaceae. Another common type is a central medulla or pith in the 59–56 range with a 'more acid' epidermis; this is common among Compositae. Some duplicate observations were made to compare flowering stems with vegetative stems, stems with leaves, and plants under different conditions or from different sources: these are grouped under the heading of "Pairs."

The system used included taking sections from the upper, middle, and lower parts of young flowering stems or occasionally young vegetative stems; or from leaves. Usually the same pH range was indicated in each part of any one stem, but variations were indicated (SMALL, 1929) by the order of the letter symbols, and here by the order of the ranges: odd cells or wall virages in a different pH range are indicated here by brackets preceded by "with." Where there are two or three pH ranges for upper, middle and lower parts, they are given in that order for each tissue: "and" means in a lower part of the stem unless there are brackets and "with" before them, indicating scattered cells.

The first series was done without BAN and BCG to break up the 52–40 range, and therefore did not differentiate cells into the ranges 52–48 or 52–44 from those in the 44–44, 44–40, and 40–40 ranges, but 52–40 is a 'more acid' range.

Abbreviations

Fs. = young flowering stem
Vs. = young vegetative stem
succ. = succulent stem or leaf
indet. = indeterminate

For tissues in stems:—

ep. = epidermis
co. = cortex
pe. = pericycle
xy. = xylem
mr. = medullary rays
coll. = collenchyma
sep. = sub-epidermis
en. = endodermis
phl. = phloem
m. = medulla
mc. = medulla, central
( ) = scattered cells

For tissues in leaves:—

ue. = upper epidermis
pp. = palisade parenchyma
fb. = fibres
gc. = guard cells of stomata
le. = lower epidermis
sp. = spongy parenchyma
bs. = bundle sheath

I. Species with epidermis < 52

A. Epidermis with cell-contents 40–40:

Puya sp., leaf; 40–40 in all tissues, except phl. 40–40 and 44–44; cuticle < 34; glands 59–59.

Crassula rosea, leaf, succ.; 40–40 ue., le., gc.; 40–40 and 52–48 pp., sp.; phl. 52–48; xy. 44–44.

Crassula portulaca, leaf, succ.; 40–40 only in ue.; xy. 44–44; 52–48 in all other tissues.

Echeveria glauca, leaf, succ.; 40–40 only in ue.; xy. 44–44; 52–48 in all other tissues.

Echinocactus ingens, Vs., succ.; 40–40 only in ep.; 52–48 sep., co., phl., mr., mc.; 44–44 xy.; en., pe., indet.: spine < 34.

Cereus flagelliformis, Vs., succ.; 40–40 and 52–48 ep.; 52–48 and 40–40 co.; 52–48 sep., en., pe., mr., mc.; 44–44 phl.; xy.; branches < 34.

Crassula sp., Vs., succ.; 40–40 and 52–48 ep., sep., co.; 52–48 in all other tissues except xy. 44–44.

Mesembryanthemum stelligerum, leaf; 40–40 and 52–48 ue., le., pp., sp., and gc.; 52–48 and 44–44 phl.; 44–44 xy; 'hairs 52–48.

The first four leaves and first stem are the only records of a completely 'very acid' (40–40) epidermis. The other three records have 'more-acid' cells (52–48) mixed with 'very acid' cells. *Puya* leaf is very-acid throughout but is peculiar in having glands which are only slightly acid, about pH 5.9. The others are members of Crassulaceae (4), Cactaceae (2) and Aizoaceae (1).

B. Epidermis with cell-contents 44–40:

Hypericum dubium (quadrangulum), Fs.; 44–40 ep., sep., co., phl.; 52–48 en., mr., mc.; xy. 44–40 and < 34.

Daucus carota, Fs.; 44–40 ep., sep., xy.; 52–48 and 59–59 co.; 59–59 en., phl., mc.; 59–59 with (56–56 and 44–40) mr.; xy. 44–40; hairs 52–48.

Inula helenium, Fs.: 44–40 ep., sep., pe., xy.; 59–59 and 40–40 co.; 59–59 en., mc.; 56–56 and 44–40 phl.; 52–48 and 56–56 mr.; hairs 52–48.

Helianthus tuberosus, Fs.; 44–40 ep.; 52–48 sep.; 59–59 co., pe., mr., mc.; 56–56 and 59–59 phl.; 44–40 and < 34 xy.; hairs 56–56 and > 62.

Polygonum aviculare, Fs.; 44–40 and 52–48 ep., sep., co., en., phl.; 44–40 and < 34 pe., xy.; 52–48 and 44–40 mr.; 52–48 and 56–56 mc.

These are not succulents, and the families are various, including Compositae (2), Umbelliferae (1), Polygonaceae (1), Guttiferae (1). All except *Hypericum* have higher pH ranges (above 52–40) in some major tissues, although this occurred in *Polygonum* only in mc. (the central medulla) in the lower part of the stem.

C. Epidermis with cell-contents 52–40:

(a) 52–40 in all tissues

Some species in the first series were examined without the use of BAN and BCG to break up the 52–40 range and were not re-examined in the second series; where differentiation was found with any tissues at higher pH ranges such species are grouped with those done later. But where no differentiation was seen, with all tissues in the 52–40 range, it might be suspected that the use of alcoholic indicators resulted in diffusion of cell-sap. The fact that considerable differentiation was found in many species with these same indicators tends to indicate that there was differentiation of pH values only *within* the range 52–40 which was not seen because the appropriate indicators were not used. These species with 52–40 recorded for all tissues are as follows:—

Orchis maculata, Fs.
Rumex acetosella, Fs.
Chenopodium album, Fs.
Saxifraga umbrosa, Fs.
Tellima grandiflora, Fs.
Ribes sanguineum, Fs.
Potentilla palustris, Fs.
Ulmaria palustris, Fs.
Geranium dissectum, Fs.
Geranium sylvaticum, Fs.
Erodium cicutarium, Fs.
Ampelopsis hederacea, Vs.
Euphorbia helioscopia, Fs.
Primula vulgaris, Fs.
Calystegium sepium, Vs.
Mentha viridis, Fs.
Calendula officinalis, Fs.
Cirsium vulgare, Fs.

The 'all-acid' families are well represented here, such as Polygonaceae (1), Saxifragaceae (2), Ribesiaceae (1), Rosaceae (2), and Geraniaceae (3), or 9 out of 18.

(b) 52–40 in epidermis and cortex:

Rumex acetosa, Fs.; 52–40 in all tissues except mr. 52–40 and 59–56, also mc. 59–56 and 52–40.

Trollius europaeus, Fs.; 52–40 in all tissues except sep. 59–56 and 52–40.

Nasturtium officinale, Fs.; 52–40 ep., sep., co., en., xy.; 56–56 and 52–40 phl.; 56–56 mr., mc.

Reseda luteola, Fs.; 52–40 in all tissues except en. 52–40 and 56–56.

Ribes grossularia, Fs.; 52–40 in all tissues except mr. 52–40 and 56–56; mc. 56–56 and 52–40.

Mercurialis annua, Fs.; 52–40 ep., sep., co., phl., xy.; 59–56 en., mr., mc.; pe. indet.

Aethusa cynapium, Fs.; 52–40 ep., sep., co., phl., xy.; 59–56 en., mc.; 52–40 and 59–56 mr.; pe. indet.

Vinca major, Vs.; 52–40 ep., sep., co., xy.; 59–56 and 52–40 en., phl.; 52–40 and 59–56 pe.; 59–56 mr., mc.

Veronica chamaedrys, Fs.; 52–40 in all tissues except phl. 52–40 and 59–56; co. indet.

Calceolaria sp., Vs.; 52–40 in all tissues except 52–40 with (59–59) mr., mc.; pe. indet.

Sherardia arvensis, Fs.; 52–40 in all tissues except sep. 52–40 and 59–56.

Tanacetum vulgare, Fs.; 52–40 in all tissues except 52–40 and 59–56 mr., mc.

This I. C. (b) group is very varied in the families represented, with only one species from most of these.

(c) 52–40 epidermis with 52–40 and 59–56 in cortex

Beta maritima, Fs.; 52–40 ep., phl., xy.; 59–56 and 52–40 sep., co.; 52–40 and 59–56 en., pe.; 59–56 mr., mc.

Brassica campestris, Fs.; 52–40 in all tissues except co. 52–40 and 59–56; en., mc. indet.

Reseda odorata, Fs.; 52–40 ep., xy.; 52–40 and 59–56 sep., co., en.; 59–56 and 52–40 pe.; 59–56 mr., mc.

Geum rivale, Fs.; 52–40 in all tissues except co. 52–40 and 56–56.

Trifolium repens, Fs.; 52–40 ep., sep., pe., phl., xy.; 52–40 and 56–56 co., en., mr.; 56–56 and 52–40 mc.

Acer pseudoplatanus, hypocotyl of seedling; 52–40 ep.. sep., en., phl., xy.; 52–40 and 59–56 co.; 59–56 and 52–40 pe., mr., mc.

Anthriscus sylvestris, Fs.; 52–40 ep., xy.; 52–40 and 59–56 sep., co., en.; 59–56 and 52–40 phl.; pe. indet.

Lysimachia nemorum, Vs.; 52–40 in all tissues except co., mr., 52–40 and 59–56; mc. 59–56 and 52–40.

Borago officinalis, Fs.; 52–40 in all tissues except co. 59–56, 56–56, and 52–40; 52–40 and 59–56 mr.; en. indet.

In this group I. C. (c) there is no sign of a taxonomic grouping, only one species each from nine different families.

(d) 52–40 epidermis with 59–56 cortex

This is one of the commoner differentiations, a more acid skin or epidermis over a slightly acid interior, especially the cortex.

Fagus sylvatica, hypocotyl of seedling; 52–40 ep., sep.; 59–56 co., mr., mc.; 52–40 also in pe., phl., xy.; en. 59–56 and 52–40.

Atriplex, sp., Vs.; 52–40 ep., pe., xy.; 56–56 sep., co., en., mr., mc.; phl. indet.

Salsola kali, Vs., succ.; 52–40 ep., xy.; 59–56 sep., co., en., mr., mc.; 59—56 and 52–40 phl.

Ranunculus auricomis, Fs.; 52–40 ep., xy.; 59–56 in all other tissues.

Ranunculus flammula, Fs.; 52–40 ep., pe., phl., xy., mr.; 59–56 sep., co.; mc. indet.

Arabis sp., Vs.; 52–40 ep., pe., xy., mr.; 59–56 co., phl.; 59–56 and 52–40 en.

Lupinus albus, Fs.; 52–40 ep., sep., en., pe., phl., xy.; 56–56 co., mr., mc.

Pisum sativum, Fs.; 52–40 ep., pe., xy.; 56–56 sep., co., en., mr., mc.; phl. 59–56 and 52–40.

Lathyrus macrorrhizus, Fs.; 52–40 ep., pe., phl., xy.; 59–56 mc.; 59–56 and 52–40 mr.; indet. sep., co., en.

Polygala vulgaris, Fs.; 52–40 ep., xy.; 59–56 co., en., phl., mr.; indet. sep., pe., mc.

Clarkia elegans, Fs.; 52–40 ep., sep., phl.; xy.; 59–56 and 52–40 co., en.; 59–56 mr., mc.

Carum petroselinum, Fs.; 52–40 ep., phl., xy., mr.; 59–56 sep., co., en., pe.

Aegopodium podagraria, Fs.; 52–40 ep., pe., phl., xy.; 59–56 co.; 59–56 and 52–40 en.; 52–40 and 59–56 mr.; sep. indet.

Oenanthe crocata, Fs.; 52–40 ep., sep., pe., phl., xy.; 59–56 co., en.; > 62 and 59–56 mr., mc.

Symphytum officinale, Fs.; 52–40 ep., sep., xy.; 59–56 co., mr., mc.; 59–56 and 52–40 en., pe.; 52–40 and 59–56 phl.

Teucrium scorodonia, Vs.; 52–40 ep., pe.; 59–56 sep., co., phl.; > 62 and 59–56 mr., mc., and with (52–40) in en.; 52–40 with (> 62) in xy.

Solanum nigrum, Vs., 52–40 ep., phl., xy.; 59–56 sep., en.; 59–59 and 59–56 co.; 52–40 and 59–56 mr., mc.; pe. indet.

Digitalis purpurea, Fs.; 52–40 ep., sep., pe., phl., xy.; 59–56 co., mr., mc.; indet. en.

Asperula odorata, Vs.; 52–40 ep., xy.; 59–56 sep., co., pe.; 59–56 and 50–40 en., mr., mc.; 52–40 and 59–56 phl.

Lonicera periclymenum, Fs.; 52–40 ep., pe., phl., xy.; 59–56 and 52–40 mr.; 59–56 mc.; indet. sep., co., en.

Campanula rotundifolia, Fs.; 52–40 ep., phl., xy.; 59–56 sep., co., mr., mc.; 52–40 and 56–56 en.; pe. indet.

Hieracium anglicum, Fs.; 52–40 ep., pe., phl., xy., mr.; 59-56 sep., co.; 59–56 and 52–40 en.; indet. mc.

Hieracium murorum, Fs.; 52–40 ep., pe., phl., xy., mr.; 59–56 sep., co., en.; 52–40 and 59–56 mc.

In this group, I. C. (d), there are a few cases of taxonomic affinities such as—2 Chenopodiaceae, 2 *Ranunculus,* 3 Leguminosae, 3 Umbelliferae, 2 *Hieracium*; but all the others are single representatives of their respective families. The 52–40 range in all these 23 species is undivided by BAN and BCG, so that a large amount of differentiation can be seen without these two indicators, when cell-contents of some cells are actually above the 52–40 pH range.

(e) 52–48 epidermis; other tissues 52–40

This group of acid plants has the epidermis cell-contents in the 52–40 range, mainly 52–48, with most other cells in the 52–40 range. These species are Polygonaceae, or succulents in one of six other families.

Haworthia reinhardtia, leaf, succ.; 52–48 in all tissues except xy. and fb. 44–44 with walls < 34.

Rumex crispus, Fs.; 52–48 ep.; 52–48 and 44–40 mr., mc.; 44–40 in all other tissues, with < 34 in pe. and xy. in part.

Rumex obtusifolius, Fs.; 52–48 ep., en., phl.; 52–48 and 56–56 sep., co., mr., mc.; pe. 52–48, 44–40, and (> 34); xy. 44–40 and < 34; coll. 44–40.

Polygonum amphibium, Fs; 52–48 ep., mc.; 52–48 and 44–40 sep., co., en., and in mr. with some < 34; 44–40 pe., phl., xy., and glands.

Polygonum persicaria, Fs.; 52–48 ep.; 52–48 and 44–40 sep., co., en., mr.; 44–40 pe., phl., xy.; 52–48, 56–56, and 44–40 mc.

Mesembryanthem tigrinum, leaf succ.; 52–48 in all tissues except xy. 44–44; prickles 52–48.

Geranium robertianum, Fs.; 52–48 and 44–40 ep., sep.; 52–48 with (44–40 and < 34) in pe., phl.; 52–48 and 56–56 mr., mc.; 44–40 xy.; hairs 52–48.

Euphorbia globosa, Vs., succ.; 52–48 in all tissues except xy. and cuticle 44–44.

Phyllocactus sp., Vs., succ.; 52–48 ep., sep., co., mr., mc.; 40–40 and > 34, with (52–48) en., and with (44–44) xy.; phl. indet., cuticle 52–48 and 40–40.

Kleinia articulata, Vs., succ.; 52–48 in all tissues except xy. 44–40; pe. indet.:

also leaf, succ.; 52–48 in all tissues except xy. 44–44.

(f) Epidermis in 52–40 range, mainly 52–48; with some major tissues 59–56

Crepis biennis, Fs.; 52–48 and 44–40 ep., sep.; 56–56 and 59–59 co., en., pe., phl.; 59–56 and 44–40 mr.; xy. 44–40 and < 34; hairs 52–48.

Taraxacum officinale, Fs.; 52–48 and 44–40 ep., sep.; 56–56 and 59–59 co., en., mr.; 56–56 with (52–48) pe., phl.; 56–56 mc.; 44–40 xy.

Agrimonia eupatoria, Fs.: 52–48 with (44–40 and < 34) ep., sep.; 52–48 and 56–56 co., mr., mc.; 56–56 and 52–48 en.; 56–56 pe., phl.; 44–40 xy.; glands 52–48: hairs 56–56.

Solanum dulcamara, Vs.; 52–48 ep.; sep. nat. pink; 59–59 co., en., pe., mr., mc.; 59–59 with (44–40) phl.; 44–40 xy.; hairs 52–48 and > 62.

Fs.; 52–40 in all tissues, differentiation not seen.

Solidago sp., Fs.; 52–48 ep., sep., and hairs; 52–48 and 56–56 co.; 56–56 en., phl.; 56–56 and 52–48 pe.; 59–59 mr., mc.; xy. 44–40.

Chrysanthemum maximum, Fs.; 52–48 ep., sep., phl.; 59–59 mc.; 59–59 with (56–56) co., en.; 52–48 with (44–40) pe.; 52–48 and 56–56 mr.; 44–40 xy.

Doronicum pardalianches, Fs.; 52–48 ep., sep.; 59–59 co., en., phl., mc.; 59–59 with (44–40) mr., and with (44–40 and < 34) pe.; xy. 44—40; hairs 52–48:

also young flower-head; 59–59 and 52–48 ep., sep.; 59–59 co., en., pe., mr., mc.; 56–56 and 59–59 phl.; 44–40 xy.

In this group five out of the seven species are Compositae.

II. Species with epidermal cell-contents in 52–40 or 59–56 ranges in different parts or mixed in the same zone, of the same stem

This larger group, like I. C. (f), includes an extra proportion of Compositae. It consists mainly of cases in which the upper, middle, or lower part of the stem sectioned differed distinctly from the other parts of these young flowering stems. As in the previous records the pH range for the lower part comes last.

Triglochin palustre, Fs.; 52–40 and 59–56 in ep., sep.; 52–40 pe., phl., xy.; 59–56 co., en.

Tradescantia virginica, Fs.; 56–56 and 52–48 in ep., sep.; 59–59 co., en., mr.; 59–56 mc.; 56–56 phl., pe.; 56–56 with (44–40) pe.; xy. 44–40; coll. 52–48.

Listera ovata, Fs.; 52–40 and 59–56 ep., sep.; 52–40 pe., phl., xy.; 59–56 en.

Stellaria holostea, Fs.; 52–40 and 56–56 ep., en., mr.; 59–56 and 52–40 pe., phl.; 52–40 sep., co., xy.

Cerastium tomentosum, Fs.; 59–59 with (52–40) ep., sep., co., pe.; 59–59 en., pe., mr., mc.; xy. 44–40; hairs 52–48.

Ranunculus bulbosus, Fs.; 52–40 and 59–56 ep., pe.; 59–56 and 52–40 phl.; 52–40 co., xy.; 59–56 en., mr., mc.

Delphinium ajacis, Fs.; 52–40 and 59–56 ep., sep., phl.; 59–56 co., en., mr.; xy. 52–40; pe. 59–56 and 52–40.

Papaver somniferum, Fs.; 59–59 with (44–40) ep., sep.;. 59–59 in all other tissues except xy. 44–40.

Papaver (Shirley), Fs.; 59–59 and 52–48 ep., sep., co., en.; 59–59 and 56–56 phl.; 59–59 and (44–40 and < 34); xy. 44–40.

Fumaria officinalis, Fs.; 52–40 and 59–56 ep., sep., co.; 59–56 and 52–40 en.; 52–40 pe., phl., xy.; 59–56 mr., mc.

Barbarea vulgaris, Fs.; 59–56 and 52–40 ep., sep., pe., mr.; 52–40 and 59–56 co., phl.; 59–56 en., mc.; xy. 44–40.

Cakile maritima, Fs.; 52–40 and 59–56 ep., sep.; 59–56 co.; 52–40 in all other tissues.

Vicia sepium, Fs.; 52–40 and 59–56 ep.; 59–56 sep., co., en., mr., mc.; 59–56 and 52–40 pe.; phl. > 62 and 59–56; xy. 52–40.

Primula auricula, Fs.; 59–56 and 52–40 ep.; 52–40 pe., xy.; all other tissues 59–56.

Buddleia variabilis, Fs.; 52–40 and 56–56 ep., en.; 59–59 and 56–56 sep., mc.; 59–56 co., mr.; 52–40 pe., phl., xy.

Valerianella olitoria, Vs.; 52–40 and 56–56 ep., sep.; 59–56 all other tissues except xy. 52–40.

Centranthus ruber, Fs.; 59–59 with (52–48) ep., sep.; 59–59 and 56–56 co.; 59–59 all other tissues except xy. 44–40.

Aster tripolium, Fs.; 52–48 and 59–56 ep., sep.; 59–59 co., en., phl., mc.; 44–40 and < 34 pe.; 59–56 and 44–40 mr.; 44–40 xy.

Aster sp. (cult.), Fs.; 56–56 and 52–48 ep., en.; 56–56 and 59–59 sep., co.; 56–56 and 44–40 pe.; 56–56 phl.; 59–59 mr., mc.; xy. 44–40; hairs 52–48 and 56–56.

Bellis perennis, Vs.; 52–40 and 59–56 ep.; 56–56 and 59–59 sep.; 59–56 co., en.; 59–59 and 44–40 pe.; 59–59 and 56–56 phl.; 56–56 with (52–40) mr., 59–59 mc. and hairs; xy. 44–40.

Spiranthes sp., Fs.; 56–56 with (44–40 and < 34) ep.; 52–48 with (44–40 and < 34) sep.; 56–56 with (< 34) co.; 56–56 en., phl.; 59–56 mr.; 59–59 mr.; 44–40 and < 34 pe.; xy. 44–40; hairs 56–56 and > 62.

Senecio jacobaea, Fs.; 56–56 with (52–48) ep., sep.; 59–59 with (52–48) co.; 59–59 en., phl., mc.; 59–59 with (44–40) pe., mr.; 44–40 and <34 xy.; 52–48 coll., cuticle, and hairs.

Crepis virens, Fs.; 59–56 and 52–40 ep., en.; 52–40 sep., co., pe., phl., xy.; 52–40 and 56–56 mr., mc.

Hieracium pilosella, Vs.; 52–40 and 56–56 ep., sep., en., phl., mr.; 52–40 and 59–56 co.; 59–56 and 52–40 mc.; 52–40 pe., xy.

More than a quarter of this group (7/24) are Compositae, with 2 Caryophyllaceae, 2 Ranunculaceae, 2 *Papaver,* 2 Cruciferae, 2 Valerianaceae, and one each from seven other families. There seems to be no regular arrangement in the position of the more acid epidermis which may occur either on the higher younger or the lower older part of the stems.

There can be all the differentiation recorded, *e. g.* for *Bellis perennis,* and yet the 'juice' may be expressed and a second decimal place accuracy claimed for a subsequent estimation of that mixed juice!

III. Species with epidermis >52

A. Epidermis with cell contents 56—56

Aloe variegata, leaf, succ.; 56–56 ue., le., pp., phl., gc.; 56–56 and 52–48 sp., xy. 44–44.

Atriplex angustifolia, Fs.; 56–56 ep., sep., co., phl.; 59–56 mc.; 59–56 with (52–40 mr.); en. 59–59 and 59–56 with (52–40); xy. 52–40; pe. indet.

Cerastium triviale, Fs.; 56–56 ep.; 59–59 sep., co., phl., mr.; 59–59 and 56–56 en.; 59–59 with (44–40) pe.; xy. 44–40; hairs 52–48.

Ranunculus acris, Fs.; 56–56 ep., sep., phl.; 59–59 and 56–56 co., en.; pe. 59–59 with (44–40 and <34); mr. 59–59 with (52–48 and 56–56); 59–59 mc.; xy. 44–40; hairs 52–48.

Ranunculus repens, Fs.; 56–56 ep.; 59–59 sep., co., en., mr., mc.; 59–59 and 56–56 phl.; 59–59 with (44–40) pe.; xy. 44–40.

Capsella bursa-pastoris, Fs.; 56–56 ep.; 59–59 with (44–40) mr.; 59–59 in all other tissues and the glands, except xy. 44–40.

Reseda suffruticulosa, Fs.; 56–56 ep., sep.; 59–59 co., en., mr., mc.; 59–59 with (44–40) phl.; xy. 44–40 and <3.4; hairs 52–48.

Viola cornuta, Fs.; 56–56 ep.; 56–56 and 59–59 sep., phl.; 59–59 co., en., mr.; pe. 56–56 and 44–40 with (59–59); xy. 44–40.

Stachys palustris, Fs.; 56–56 ep., sep., co., en., mc.; 52–48 and 56–56 phl.; 52–48 with (44–40 and <34) pe.; 59–59 with (44–40 and <34) mr.; xy. 44–40 and <34; hairs 52–48.

Lamium album, Fs.; 56–56 in all tissues except xy. 44–40, and hairs 52–48 and 56–56.

Veronica beccabunga, Vs.; 56–56 ep., sep., co.; 59–56 and 52–40 en.; 52–40 and 59–56 phl.; 56–56 and 59–59 mr.; xy. 44–40; pe., mc., indet.

Digitalis sp. (white, cult.), Fs.; 56–56 ep.; 56–56 and 59–59 sep., co., en.; 59–59 mr., mc.; pe. 56–56 and 59–59 with (44–40 and <34); phl. 52–48 and 59–59; xy. 44–40; hairs 52–48.

Galium verum, Vs.; 56–56 ep., sep., co., mr., mc.; 52–48 en. and hairs; 52–48 and 56–56 phl.; xy. 44–40 and <34; pe. indet.

Scabiosa arvensis, Fs.; 56–56 ep., sep.; 59–59 and 56–56 co., en., pe., phl.; 59–59 mc.; 56–56 and 44–40 with (59–59) mr.; 44–10 with (<34) xy.; hairs 52–48.

Scabiosa succisa, Fs.; 56–56 ep., sep., en., pe., phl.; 56–56 and 59–59 co.; 59–59 mc.; 56–56 with (44–40) mr.; xy. 44–40 and <34.

Senecio sylvaticus, Fs.; 56–56 ep., sep.; 59–59 co., en., phl., mr., mc.; 44–40 and <34 with (59–59) pe; xy. 44–10; hairs 52–48.

Centaurea montana, Fs.; 56–56 ep., sep., co., en., phl., mr., and hairs; 44–40 and <34 with (52–48) pe.; 59–59 mc.; xy. 44–10.

Lapsana communis, Fs.; 56–56 ep.; 56–56 and 59–59 sep., co.; en.; 59–59 pe., phl., mc.; 59–59 with (44–40) mr.; xy. 44–40; cuticle 52–48.

Hieracium aurantiacum, Fs.; 56–56 ep., sep.; 59–59 with (56–56) co., en., phl.; 59–59 pe., mc.; 59–59 and 56–56 mr.; xy. 44–40; hairs 52–48 and 56–56; cuticle 52–48.

In this group, III. A., there are 4 Compositae, 2 each of Ranunculaceae, Labiatae, Scrophulariaceae, and Dipsaceae; and one each of seven other families, not one of which is a member of the previously enumerated "all acid" families. With the exception of the usual more acid xylem and some pericycles, the inner tissues of these III. A. stems are also in the 59–56 range with the closely similar ranges, 59–59, 59–56, 56–56, differentiated in various ways, and quite frequently with scattered more acid cells in these less acid tissues. In the stems the xylem is, as usual, almost uniformly in the 44–40 range; the range 52–40 for the xylem in other groups occurs mainly when that range has not been subdivided by the use of BAN and BCG.

B. Epidermis with cell-contents 56 and 59, or 59–56

56 and 59

Alisma plantago, Fs.; 56–56 and 59–59 ep., sep., co., en., mc., and unlignified xylem; 52–48 with (44–40) phl.; 52–48 with (44–40 and <34) pe.

Stellaria media, Fs.; 56–56 with (59–59) ep.; 59–59 and 56–56 en., pe.; 59–59 sep., co., phl., mr.; mc. indet.; xy. 44–40; hairs 56–56.

Silene maritima, Fs.; 56–56 and 59–59 ep., sep., co., en., phl., mr.; 59–59 mc.; pe. 52–48, 44–40, <34; xy. 44–40.

Trifolium pratense, Fs.; 56–56 and 59–59 ep., sep., co., en., phl.; 59–59 mr., mc.; pe. 52–48 and 44–40; xy. 44–40; hairs 56–56.

Salvia officinalis, Vs.; 56–56 and 59–59 ep., sep., co., en., phl.; 56–56 mr., mc., and hairs; 44–40 pe., xy.

also Fs.; 52–40 in all tissues, differentiation not seen.

59–56

Convallaria majalis, Fs.; 59–56 in all tissues except phl. 52–40 and 59–56; also pe., xy. 52–40.

Orchis mascula, Fs.; 59–56 in all tissues except mr., mc. 59–56 and 56–56; also pe., xy. 52–40.

Urtica dioica, young Vs.; 59–56 ep., sep., mc.; 52–40 and 59–56 co., en., phl., mr.; pe., xy. 52–40.

Papaver rhoeas, Fs.; 59–56 in all tissues except sep. 59–56 and 52–40; also xy. 52–40, and large hairs < 34.

Cheiranthus cheiri, Fs.; 59–56 ep., sep., co., phl.; 52–40 and 59–56 mr.; 52–40 mc., xy.; en., pe. indet.

Cardamine pratensis, Fs.; 59–56 in all tissues except mr. 59–56 and 52–40; and xy. 52–40.

Hedera helix, hypocotyl of seedling; 59–56 in all tissues, including unlignified xylem; no differentiation seen.

Nicotiana tabacum, Vs.; 59–56 in all tissues except xy. 52–40; sep., en. indet.

Pedicularis palustris, Fs.; 59–56 ep., mr., mc.; 59–56 and 52–40 en.; 52–40 pe., phl., xy.; sep., co. indet.

Chrysanthemum hort., Vs.; 59–56 in all tissues except 52–40 pe., xy.

Senecio vulgaris, Vs.; 59–56 in all tissues except xy. 52–40; pe. indet.

Centaurea nigra, Fs.; 59–56 ep., sep., co.; 52–40 en., pe., mr., xy.; indet. phl., mc.

59 and 56

Gasteria verrucosa, leaf, succ.; 59–59 and 56–56 ue., le., gc.; 56–56 pp., sp., phl.; xy. 44–44; cuticle < 34.

Phlox, sp., cult., Fs.; 59–59 and 56–56 ep., sep.; 59–59 co., phl., mc.; 59–59 with (52–48) en., mr.; 59–59 with (44–40 and < 34) pe.; xy. 44–40 and < 34.

In this group, III. B., there is a wide representation of families, 3 Compositae, 2 Caryophyllaceae, 2 Cruciferae, and one each from twelve other families.

C. Epidermis with cell-contents 59–59

Arenaria peploides, Fs.; 59–59 ep., sep., co., mc.; 59–59 and 56–56 en.; 59–59 and 44–40 mr.; 44–40 and < 34 pe.; xy. 44–40.

Heracleum sphondylium, Fs.; 59–59 ep., sep., co., en., pe., mc.; 59–59 and 56–56 phl.; 59–59 with (56–56 and < 34) mr.; xy. 44–40; hairs 52–48 and 56–56.

Stachys sylvatica, Fs.; 59–59 ep., sep., co., en., phl., mc.; 59–59 with (44–40 and < 34) pe., mr.; xy. 44–40; hairs 52–48 and 44–40.

Lycopersicon esculenta, Vs.; 59–59 in all tissues except xy. 44–40; hairs 52–48 and 56–56.

Galium aparine, Vs.; 59–59 in all tissues except xy. 44–40; and hairs 52–48; pe. indet.

Sonchus oleraceus, Fs.; 59–59 ep., sep., co., en., mr., mc.; 59–59 with (44–40) pe.; 59–59 and 56–56 phl.; coll. and xy. 44–40; hairs 52–48.

This III. C. group is very small, only six species from six different families with the epidermal cell-contents of no greater acidity than pH 5.9. This emphasises the large number of species in class I with its groups ABC and all their subdivisions: a "more acid" skin, either complete or in parts is common among the plants examined in this series.

IV. Species with epidermal pH undetermined

Where the indicators for various reasons did not yield a pH range, the stems concerned could not be classified into any of the above groups, but the records for their other tissues are included below for these six species.

Matthiola incana, Fs.; ep. and pe. indet.; 59–56 sep., co., en., phl.; 52–40 mr., xy.

Iberis amara, Fs.; ep., pe., phl. indet.; 59–56 sep., co., en., mc.; 52–40 and 59–56 mr.; xy. 52–40.

Ligustrum vulgare, Vs.; ep., sep. indet.; 59–56 co., phl.; 59–56 and 52–40 en., pe., mr., mc.; xy. 52–40; but see later details.

Ajuga reptans, Vs.; ep., sep., pe., mc. all indet.; 59–56 co., en., phl.; 52–40 and 59–56 mr.; xy. 52–40.

Mimulus luteus, Fs.; ep. indet.; 59–56 sep., co., pe., phl., mc.; 52–40 and 59–56 en., mr.; xy. 52–40.

Artemisia sp., Vs.; ep., en., pe. indet.; 59–56 sep., co., mr., mc.; 52–40 phl., xy.

As usual, these higher pH types are not characteristic of any one family: 2 Cruciferae and 1 each of four other families. In all these cases the epidermal cell-contents could well be of a mixed character.

V. Paired records

Many of the variations in the above records show that the acidity of the cell-sap in any one tissue can vary from one part to another of the same stem, and this raised the problem of variation from one stem to another; from the young flowering stem to the young vegetative stem of the same kind of plant; and of similar stems from plants under different conditions. These are sometimes problems connected with horticulture or agriculture, but from a purely academic point of view examples of paired records from REA and SMALL's extensive list (SMALL 1929) can be used to show what happened in particular cases.

A. Duplicate records

These show that the range of variation for the same tissue in the same species is usually not large. Many other observations on selected species throughout the year or throughout their annual growth, as in sunflower and broad bean, have also demonstrated a considerable degree of "universal reproducibility" of tissue pH values, but variation with strains of plants or with external conditions are to be expected.

Examples:

Anemone japonica, Fs., 2 sets; ep., sep., co., en., phl. all 59–59 in A, and 56–56 in B; mr., mc., 59–59 in both, also xy. 44–40; pe. 44–40 and <34 in both but with scattered cells (59–59) in A, and (56–56) in B. These two, A and B, differed only within the 59–56 range.

Aconitum napellus, Fs., 2 sets; ep., sep. 56–56 and 59–59 in A, but only 56–56 in B; in both A and B phl., mr., mc. 59–59; xy. 44–40; co., en. in A 59–59; in B outer co. 59–59, inner co. 56–56, and en. 56–56 and 44–40; pe. 44–40 and < 34 in both, but with (59–59) in A, and with (56–56) in B: also Vs.; 59–56 in all tissues except pe., xy. 52–40. Here again the difference is mainly within the 59–56 range with "more acid" in the endodermis of B.

Dicentra formosa, Fs., 2 sets; ep., sep. 56–56 A, 59–59 and 56–56 B; co., en. 56–56 and 44–40 in both, but with (59–59) in B; mr., mc. 59–59 in both, but 56–56 also in B; pe. 44–40 in A, but 59–59 with (56–56 and 44–40) in B; in both, phl. 59–59; xy. 44–40.

Again there is variation within the 59–56 range, but here there is a mixture of 59 and 56 in pe. of B and not in A. The strong difference in phloem (5.9) and xylem (44–40) should be noted; this is by no means usual.

Geum sp., cult., Fs., 2 sets; ep., sep., phl., xy. 44–40 in both A and B;
A 56–56 co., en., mr., mc.; 44–40 pe.;
B 52–48 co., en.; 52–48 and 56–56 mr., mc.; 44–40 and (< 34) pe. Here B is definitely more acidic than A in co., en., mr., mc., and in some cells of pe.

Geranium sp., cult., Fs., 2 sets; A — 52–48 ep., sep., co., pe., phl.; 52–48 and 56–56 en.; 56–56 with (59–59) mr., mc.; xy. 44–40; hairs 56–56 and 59–59.

B — 44–40 ep., sep., phl., xy.; 56–56 co., en.; pe. 52–48 with (44–40 and < 34); 52–48 with (44–40) mr., mc.; hairs 52–48 and 59–59. Here B is more acidic than A in ep., sep., phl., also in mr., mc., and some hairs; but B is less acid than A in co., en.

Pelargonium sp., cult., Fs., 2 sets; A — 44–40 ep., sep., co., en., phl., xy.; 44–40 and (56–56) mr., mc.; B — 52–48 ep., sep., co., en., phl.; 56–56 and (52–48) mr., mc.; both A and B — pe. 44–40 and < 34; xy. 44–40; hairs 52–48.

Pelargonium is in the "all-acid" family, Geraniaceae, and here the pH rises above the 52–40 range only in some medullary cells of A and the main medullary tissues in B; with A rather more acidic (44–40) than B (52–48) in most of their tissues.

Limnanthes douglasii, Fs., 2 sets; A — 59–56 in all tissues, except xy. 52–40; B — 59–59 ep., sep., pe., mr.; 59–59 with (56–56) co., en., phl., mc.; xy. 44–40. *Limnanthes* was in the Geraniaceae of Bentham and Hooker, but it is now removed as Limnanthaceae from the Geraniales to the Sapindales of Engler's classification, on the position or posture of the ovule. The difference between the 2 sets of records, A and B, is within the 59–56 range, and the higher pH range of the tissues is an interesting confirmation of the removal of this genus from the otherwise almost "all acid" Geraniaceae.

Armeria maritima, Fs., 2 sets; A — 52–48 ep., sep., co., en.; pe. 52–48 with (44–40 and < 34); phl. 52–48 and 44–40; mr. 44–40 and 52–48; mc. 56–56; xy. 44–40; hairs 52–48.

B — 52–48 and 56–56 ep., sep., co., en.; pe. 56–56 with (44–40 and < 34); 56–56 phl., mr., mc.; xy. 44–40. The upper part of these two was almost the same 52–48 but the lower part of A remained 52–48 while that of B was 56–56 in ep., sep., co., en., with the same higher range for phl., mr., and mc.

Myosotis vulgaris, Fs., 2 sets; A and B both, 56–56 ep., phl.; 59–59 mc.; 44–40 xy.; 56–56 with (44–40 and < 34) pe.; hairs 56–56:

A — 56–56 sep., en.; 52–48 and 56–56 co.; 52–48 and 59–59 mr.;

B — 59–59 sep.; 59–59 with (44–40) co.; 56–56 and 59–59 en.; 44–40 and 52–48 mr. Apart from the "more acid" parts of B (co. and mr.) these two were very similar within the range 59–56.

Prunella vulgaris, Fs., 2 sets; A and B ep. indet.; 56–56 sep., phl.; xy. 44–40.

A — 59–59 co.; 56–56 and 44–40 en., pe.; 56–56 mc.; 56–56 with (52–48 and 44–40) mr.;

B — 56–56 and 59–59 co., mc.; 56–56 and 52–48 en.; 56–56 pe., phl., mr., and hairs.

Here B was below 52 only in xy., and in en. of the lower part of the stem; while A was "more acid" in en., pe., xy. in the lower part, also in mr., xy., in all parts of the A stem.

Solanum tuberosum, Vs., 2 sets; A — 59–59 in all tissues except xy. 44–40, and hairs 52–48 and > 62.

B — 52–48 and 56–56 ep.; 59–59 and 56–56 sep.; 59–59 with (56–56) co.; 56–56 and 59–59 en., phl.; 56–56 and 44–40 pe.; 59–59 and 44–40 mr.; 59–59 mc.; A and B xy. 44–40, and hairs 52–48 and > 62.

A was very uniform, and B varied from A mainly within the 59–56 range, with the "more acid" skin in the top, and "more acid" pe., mr., in the lower part of the stem. The main storage tissue of the tuber is known to be about pH 6.2 when the intercellular carbon dioxide has escaped, and the main stem tissues are reported as 59–59 by Small (1929, p. 270/71) with carbon dioxide of respiration bringing the cell-sap pH down to 56–56 or even to 52–48: but the 44–40 range in pe. and mr. indicates extra organic acid in these tissues in stem B above, similar to that in the "more acid" epidermis which covers the whole of the "eyes" of the tuber and only the bases of aerial stems of the potato. Small (1929, pp. 265–273) reported fully upon the pH ranges to be found under different conditions in various parts of the potato plant.

Plantago lanceolata, Fs., 2 sets; A and B — 59–59 co., en., mc.; 44–40 xy.:

A — 52–48 and 56–56 ep., sep.; 44–40 pe., phl.; 59–59 with (52–48 and 56–56) mr.; hairs 44–40;

B — 56–56 ep., sep., phl.; 56–56 and 44–40 pe.; 59–59 mr.

Again the variation from A to B was mainly in the 59–56 range, and probably subject to carbon dioxide effects as in the potato; but there was a "more acid" epidermis in the top part of A, and a less acid pericycle in the top part of B, also scattered acid cells in the 59–59 medullary ray region of A but not in B.

Matricaria inodora, Fs., 2 sets; A and B — 59–59 co., mc.; 44–40 xy.;

A — 52–48 and 44–40 ep., sep.; 59–59 en., mr.; 44–40 pe.; 44–40 with (59–59) phl.;

B — 56–56 and 59–59 ep., sep.; 56–56 en.; 56–56 and 44–40 pe.; 44–40 and 56–56 phl.; 59–59 and 44–40 mr.

Here both upper and lower parts of A are "more acid" than those of B in ep., sep.; while the phloem in both A and B upper parts is "more acid" (44–40), but less acid in scattered cells (A), or lower down the stem (B). On the other hand B is "more acid" than A in the medullary ray region of the lower part of the stem.

Hypochoeris radicata, Fs., 2 sets; A — 56–56 ep., sep.; 59–59 co., en., phl., mr.; 59–59 with (44–40 and <34) pe.; mc. indet.;

B — 52–48 ep., sep., and hairs; 59–59 and 56–56 co., en.; 56–56 with (44–40 and <34) pe.; 56–56 phl.; 59–59 with (44–40) mr.: A and B — 44–40 xy.

B was "more acid" in ep., sep., than A and scattered "more acid" cells in mr., but otherwise the variation in these two stems was again within the 59–56 range.

The brief notes on these 12 pairs of young flowering stems and on the young vegetative stems of potato tend to repeat the same statement that the variation found between two stems of the same kind of plant is *within* the 59–56 range with or without some tissues or odd cells in a tissue showing as "more acid" in one or other of these paired records.

B. Pairs of flowering and vegetative stems

Among the many records tabulated by Small (1929) there are some which pair as comparisons of young flowering stems with young vegetative stems which, of course, often develop later into flowering stems. The physiological age may be taken as greater in the latter than in the former, although the actual tissues being a newer development may be younger in the flowering stems. Some examples will give a factual basis for consideration of the pH ranges in this connection.

Lilium tigrinum, Fs.; 52–40 in all tissues of upper part, but in lower part — 52–40 ep., pe., phl., xy.; 59–56 sep., co., en., mr., mc., Vs.; 59–56 in all tissues of all parts, except pe., xy. 52–40.

In this case there is a strong similarity between Vs. and the lower part of Fs., in sep., co., en., mr., mc., that is in the main bulk of the parenchymatous tissue, at 59–56, as well as pericycle and xylem at 52–40. The epidermis and phloem of both upper and lower parts of Fs. are the only two tissues which are consistently more acid in Fs. than in Vs. This, at first sight, appears to be untrue, but pe. and xy. are in the same 52–40 range in all parts of both stems, and the greater acidity is confined to the *upper* part of Fs. in sep., co., en., mr., mc., which are 59–56 in the lower part of Fs.

Vicia faba, Fs.; 52–40 in all tissues; differentiation not seen. Vs.; ep. 59–56 with (> 62 and 52–40); sep. 59–56 and 52–40; 59–56 co., en., mr.; < 34 pe.; xy. 52–40.

REA and SMALL (1927 a) reported fully on the pH ranges found at various stages in plants of this species, including notes of ep. at 5.6 in a green variety with 52–48 ep. in a white variety and in occasional "green" plants, also that the epidermis was 52–48 at the tip but 56–56 lower down in mature plants (see SMALL 1929, p. 233). In this pair the higher range occurs only in Vs., which on the whole is definitely less acidic than Fs.

Tropaeolum majus, Fs.; 52–40 in all tissues; differentiation not seen. Vs.; 52–48 and 56–56 ep., sep.; 56–56 co., phl.; 56–56 and 59–59 mr., mc.; 56–56 with (44–40) pe.; xy 44–40.

Here again the vegetative stem is largely less acid than the flowering stem in many main tissues.

Epilobium montanum, Fs., 2 sets; A and B ep. natural pink; xy. 44–40.

A — 56–56 in all other tissues, also hairs.

B — 44–40 sep., phl.; 56–56 co., en., mc.; 44–40 and 52–48 mr.; hairs 52–48:

also Vs.; ep. nat. pink; 44–40 sep.; 52–48 and 56–56 co., mc.; 52–48 phl., mr.; 44–40 and 56–56 en.; 44–40 and < 34 xy.; pe. indet.; hairs 56–56.

Fs. B. was more acid than Fs. A. in sep., phl., mr., and hairs; but these differences are partially bridged by the 52–48 range which occurred in Vs.; 56–56 passing through 52–48 to 44–40 and 52–48 in the medullary rays. The main parenchymatous tissues as cortex and central pith (mc.) were 56–56 in all parts of Fs. A and B, and in the lower part of Vs., which leaves the upper youngest part of Vs. more acid than later stages, like the epidermis of the broad bean (see *Vicia faba* above).

Scutellaria galericulata, Fs.; 52–40 in all tissues; differentiation not seen. Vs.; 59–59 in all tissues except pe. 59–59 with (44–40 and < 34); xy. 44–40 and < 34; hairs 56–56.

There is, in this case, an almost complete contrast which indicates considerable acidification in the flowering stem, as a later development in the life of these plants.

Lamium purpureum, Fs.; ep. indet.; 59–59 in all tissues except xy. 44–40, and hairs 52–48.

Vs.; ep. indet.; 59–56 in all tissues except 59–56 and 52–40 sep., en.; and 52–40 xy.

There is very little significant difference between Fs. and Vs. except in the "more acid" sep. and en. in the lower part of Vs.

Antirrhinum majus, Fs.; 52–40 and 59–56 ep., sep., co.; 59–56 en., mc.; 59–56 and 52–40 mr.; 52–40 pe., phl., xy.

Vs.; 56–56 ep.; 56–56 and 59–59 sep., co., phl.; 59–59 en., pe., mr., mc.; xy. 44–40; hairs 52–48.

Here the Vs. was nearly all less acid, and the Fs. was less acid only in the medullary tissues of parts and in ep., sep. and co. of the lower part of the flowering stem.

Scrophularia nodosa, Fs.; 52–40 ep., sep., co., pe., phl.; en. > 62 and 59–59 with (52–40); 59–56 mr., mc.; xy. 52–40 with (> 62).

Vs.; 59–59 ep., sep., co., phl.; 56–56 en.; 52–48 and 56–56 mr.; 59–59 and 59–56 mc.; 44–40 and < 34 pe., xy.

Apart from the usually "more acid" pe. and xy., the outer tissues of Fs. were definitely acidic while the corresponding cell-saps of Vs. were in the 59–56 range. This is in accordance with a kind of general rule that the Fs. tends to be more acid than the Vs.

Echeveria metallica, Fs., succ.; 52–40 in all tissues; no differentiation seen.

Leaf, succ.; 40–40 and 52–48 ue., le.; 52–48 pp., phl., gc.; 52–48 with (56–56) in sp.; xy. 44–44.

Apart from the scattered less acid cells in the spongy parenchyma of the leaf, the general 52–40 of the Fs. was merely differentiated by BAN and BCG in the later investigation.

Crassula lycopodioides, Vs., succ.; 52–48 in all tissues except xy. 44–44; pe. indet.

Leaf, succ.; 52–48 in all tissues except xy. 44–44.

Here there is no difference between leaf and vegetative stem.

In these eight pairs of Fs.-Vs. comparisons the flowering stem tends to be more acidic than the vegetative stem in *Lilium, Vicia, Tropaeolum, Scutellaria, Antirrhinum,* and *Scrophularia. Lamium* showed no difference except within the 59–56 range; and *Epilobium* was rather complicated with the vegetative stem tending to be more acidic than most of the tissues of both flowering stems which, however, showed marked differences between Fs. A. and Fs. B. in subepidermis, phloem, and medullary rays: the cortex, endodermis, and central medulla of the lower part of Vs. was the same (56–56) as in all parts of both flowering stems; so that it was only the youngest part of the vegetative stem which showed a greater acidity in *Epilobium,* and this same phenomenon is noted for the epidermis in *Vicia faba.*

C. Pairs and conditions

The ideas, that very young stems tend to be more acidic, that stems become less acid as they become metabolically mature, and then change back to greater acidity when they pass to the flowering stage, are all suggested on the very slender base of what has been observed in only six species. This suggested sequence is, however, supported by only two out of the three following pairs of younger and older young flowering stems; although there is more evidence to be found among the previously quoted differences between upper and lower *parts* of young flowering stems.

Plantago major, Fs., 2 sets; A (younger); 52–40 in all tissues except sep., co. 52–40 and 59–56.

B (older); 52–40 and 59–56 ep.; 56–56 sep., co., en., phl.; 56–56 and 44–40 pe.; 59–56 mr., mc.; xy. 44–40.

Apium graveolens, Fs., 2 sets; A (younger); 52–40 in all tissues; sep., co. indet., no differentiation seen:

B (older); 56–56 ep., sep., co., pe., phl.; mr. 52–40 with 56–56 in parenchyma; xy. 52–40.

Alchemilla vulgaris, Fs., 2 sets; A (younger); 56–56 and 52–40 ep.; 56–56 sep., mc.; 52–40 and 56–56 co., pe.; 52–40 en., phl., xy., mr.

B (older); 52–40 in all tissues; no differentiation seen.

Varied conditions may affect the cell-sap pH of some tissues and not others, as may be seen in the following paired records.

Endymion (Scilla) nonscriptus, Fs., 2 sets; A wild; 59–56 in all tissues except pe., xy. 52–40.

B cult.; 56–56 in all tissues, except pe., xy. 52–40, and phl. 56–56 and 52–40.

Lychnis sp., Vs., 2 sets; A (in sun); 56–56 ep., en., phl.; 59–56 and 56–56 sep.; 56–56 and 59–56 co.; 52–40 and 56–56 pe.; 59–56 mr.; mc. indet.; xy. 52–40:

B (in shade); 52–40 ep., en., phl., xy.; 52–40 and 59–56 sep., co., mr.; also Fs.; ep., mr., mc. indet.; 59–56 sep., co., en.; 52–40 pe., phl., xy.

The upper part of the shaded Vs. was all 52–40 as compared with the plant in the sun where most of the cell-saps were within the 59–56 range which also occurred in sep., co., mr. of the lower part of the shaded plant. Shading may have produced the acidification in this case, since the subepidermis and cortex of the flowering stem were also in the 59–56 range.

Achillea millefolium, Fs., 2 sets; A (dry, sunny); 44–40 ep., sep., xy.; 52–48 co., hairs; 44–40 and 59–59 en., phl.; 44–40 and < 34 pe.; 59–59 mc.; 59–59 with (44–40) mr.;

B (shady) 56–56 ep., sep., co., en., phl.; 44–40 and < 34 with (56–56) pe.; 59–59 mc.; 59–59 and 56–56 mr.; hairs 52–48.

Contrary to the suggestion made for *Lychnis,* shading in *Achillea* is correlated with the less acid range 59–56, especially as compared with the outer tissues in the sun, ep., sep., co., which are within the 52–40 range. The medullary cell-saps in both are in the 59–56 range; and as these are inner tissues the maintenance of the higher pH values may be due to more effective buffering against carbon dioxide effects in *Achillea* than in *Lychnis.* There are available *theoretical* explanations of most contradictory pH phenomena.

Campanula sp., cult., Fs., 2 sets; A (dry habitat); (52–48 ep., sep.; 52–48 and 56–56 co.; 56–56 en., phl.; 56–56 and 59–59 mc.; 56–56 with (52–48) mr.; 56–56 and 44–40 pe.; 44–40 xy.;

B (damp habitat); 56–56 ep.; 59–59 sep., co., en., pe., mc.; 56–56 and 59–59 phl.; 59–59 and 44–40 mr., 44–40 xy.

The "more acid" ep., sep., co. of the upper part of the flowering stem from a dry habitat was the main difference between these two sets; the lower part of A had only the ep., sep. 52–48, and its pe., xy. at 44–40, so that it was like B. This *Campanula* pair is very similar to *Achillea,* and may also be well buffered against carbon dioxide effects.

Dianthus caryophyllus, Fs., 2 sets; A in medium habitat and B in damp habitat. The middle part of these two stems was 59–59 in ep., sep., co., phl., mc.; also the upper and lower parts of B were 59–59 in sep., co., en., phl., mr., mc. The slightly more acid tissues were in the upper and lower parts of A, ep., phl. 56–56; the upper part of A sep., co., 56–56, and the middle part of A en. 56–56; so that most differences were within the 59–56 range, and not very significant. The xylem was 44–40 as usual, and the pericycle was 56–56, 44–40, < 34 in A, and 59–59, 44–40, < 34 in B, so that in this tissue the actual difference was again within the 59–56 range. This reads as a complicated story but the letter notation makes it plain, thus:—

	ep.	sep.	co.	en.	pe.	phl.	xy.	mr.	mc.
A —	cac	caa	caa	cca	chk	cac	h	a	a
B —	cac	a	a	a	ahk	a	h	aea	a

Chrysanthemum segetum, Fs., 2 sets, A cult., B from seaside. A and B 44–40 and 59–59 ep., sep.; 59–59 mc.; 44–40 xy.:

A — 56–56 and 59–59 co., en.; 56–56 phl.; 59–59 with (< 34) mr.; 44–40 and < 34 pe.:

B — 59–59 co., en., phl., mr.; 59–59 with (44–40) pe.

The seaside stem was more uniformly at a high pH, 59–59; but ep. and sep. were "more acid" in the upper parts of both stems and most of the differences were within the 59–56 range.

Sedum praealtum, leaf, succ.; 2 sets, A in daylight, B after 20 hours in darkness. The letter notation seems to be clearer in this case thus:—

	ue.	pp.	sp.	xy.	phl.	gc.	le.	
A —	ce	c	c	g	eg	e	e	— light
B —	ec	e	e	g	e	ec	ec	— dark

The main change is from **c** or 56–56 to **e** 52–48 in the palisade and spongy parenchyma with darkness: the other differences are re-arrangements of the balance between **e** and **c** in the upper and lower epidermis, but the record of 56–56 in any part of le. and of guard cells in the dark when these are 52–48 in light is somewhat surprising. The changes in titratable organic acids with light and darkness have been fully reported upon by Wood (1952) who found variations in mg.-equivalent concentrations but, quite appropriately, presented no systematic pH data for "leaves crushed in a mortar and ground to a fine consistency," "Water was added and the pulp boiled for 5 minutes." Obviously any resulting pH values would not have much relation to those noted above for cell-contents of fresh sections.

VI. Summary of pH ranges for tissues

Epidermis (ep). The cell-sap of epidermal cells varies from pH 4.0 to pH 5.9, with epidermal hairs varying beyond the normal R. I. M. range at both ends, from < 34 to > 62. Greater acidity (52–40 range) is commoner than less acidity in epidermal cells of flowering stems. Apart from the 35 pairs of sets from the same species, there are 92 species with epidermis always within the 52–40 range, 44 always within the 59–56 range, 24 mixed or varying from one range to the other in different parts of the same stem, and 6 with the epidermal pH undetermined. Among the 35 paired sets 20 pairs show mixtures or variations into both ranges; only one species is always 44–40; four are consistently within the 52–40 range while 8 are always within the 59–56 range and two have not been determined. Most of these observations were made on young flowering stems, and among the young vegetative stems examined the epidermal cells were commonly in the 59–56 range in whole or in part.

Sub-epidermis (sep.). This layer is sometimes 'more acid,' or intermediate between acid epidermis and less acid cortex but more frequently it tends to be the same as the general cortex.

Cortex (co.). Apart from the 'all acid' families in which nearly all tissues are within the 52–40 range, the cortex is more commonly in the 59–56 range, always in that range when the epidermis is 59–56 as in the 44 species in section III above, (except in the upper Vs. of young nettles), and nearly always among the 39 species in section I. C. (c), (d), (f) above, where the epidermis in 52–40 with at least some of the main tissues in the 59–56 range. SMALL (1929, p. 118) gave ratios for the two pH ranges in the cortex in a number of families totalling 81 records in the 59–56 range to 18 in the 52–40 range.

Endodermis (en.). This tissue is often the same as the cortex, relatively less acid (59–56), sometimes with more acid suberised or lignified cell-walls. SMALL's list of families with ratios of 59–56 and 52–40 records *(loc. cit.)* for this tissue includes Caryophyllaceae 6/0, Ranunculaceae 6/1, Cruciferae 7/1, Umbelliferae 5/1, Compositae 22/4, and totals 76/14.

Pericycle (pe.). This is 'more acid' in all the monocotyledons examined, and in most of the dicotyledons, but less acid cells occur in the pericycle of some families *e. g.* Caryophyllaceae, Solanaceae, Valerianaceae, Dipsaceae, also in *Anemone* and *Aconitum.*

Phloem (phl.). The sieve-tube contents tend to be less acid (59–56) in some families, the ratios of records 59–56/52–40 being given by SMALL *(loc. cit.)* as in Caryophyllaceae 6/0, Ranunculaceae 5/2, Papaveraceae 4/1, Cruciferae 5/2, Labiatae 6/3, Valerianaceae 4/1, Dipsaceae 2/0, Compositae 19/9. But elsewhere the commonest pH range for sieve-tube contents is the range 52–48, with the xylem at 44–40. In some cases the sieve-tube walls could be differentiated as 52–40 with contents at 59–56; and the callus plates giving a virage indicating 62–59 or 62–62.

Xylem (xy.). This part of the stem is much differentiated structurally. The xylem parenchyma is sometimes in the 59–56 range but is often as acid as the lignified walls of vessels and tracheides when these are in the 44–40 range. Mature lignified walls especially of fibres are frequently very acid, < 34, but these have probably passed through the 44–40 stage during their lignification, as is indicated by special studies of growth stages in sunflower and broad bean. The xylem is 44–40 at an early stage, in the procambial strands of seedlings in both sunflower and broad bean.

Medullary rays (mr.). In the young stems of herbs there is no occasion to use the 'timber' terms "wood-ray" and "bast-ray." The tissue determined as mr. is that part of the medulla which lies between the successive ridges of primary xylem. These rays are often not differentiated from the central medulla or pith, either structurally or in the pH range of their cell-sap, but in some stems this ray region is more acidic than the central tissue, and sometimes it varies within the same part of the same stem. It tends to be in the 59–56 range in the same families as are mentioned above under endodermis and phloem. SMALL (1929, p. 120) gives detailed ratios which for the selected families total 46 in the 59–56 range to 11 in the 52–40 range.

Medulla, central (mc.). Apart from the stems which are 'all acid' or mostly acid, the pith tends to be in the 59–56 range. It is mostly thin-walled parenchyma which may break down and leave an air space, so that it is often noted as "indet" or undetermined. SMALL *(loc. cit.)* gives the following ratios for 59–56/52–40:— Chenopodiaceae 4/1, Caryophyllaceae 4/0, Ranunculaceae 5/1, Papaveraceae 5/0, Cruciferae 5/2, Leguminosae 5/1, Umbelliferae 5/1, Boraginaceae 2/1, Labiatae 6/0, Solanaceae 4/0, Scrophulariaceae 7/1, Valerianaceae 2/0, Dipsaceae 2/0, Compositae 22/1; totalling 78/9. So that there is a strong tendency for the pith to be only slightly acid (59–56) except in stems which are 'more acid' in all their tissues.

Hairs. These are most frequently in the 52–48 range, and sometimes 44–40, total 31 species with 'more acid' hairs; while 14 species have hairs in the 59–56 or 59–59 range. Both types occur on the same stem in eleven species.

As SMALL pointed out in 1929 (*op. cit.*), the close relationship of some of the all-acid families indicates some general physiological relationship analogous to the structural similarities. The acidity of succulents is well known, but some all-acid plants are not succulent, while the 59–56 pH range for the main tissues of *Salsola kali* indicates clearly that succulence and acidity have no exclusive causal connection. For ordinary mesophytic plants, this survey was sufficiently wide to demonstrate the usual limits and the numerous variations within these limits for the pH ranges of cell-sap among angiosperms. Detailed studies were made of sunflower and broad bean and of some special aspects: these were reported in a series of papers. The sunflower report has been summarised in Chapter I, and a summary of the broad bean report follows below.

3. R. I. M. pH ranges for *Vicia faba*

In Chapter I a brief summary is given of MARTIN's investigation of pH ranges during the development and growth of all parts of the sunflower *(Helianthus annuus)*. REA and SMALL (1927) made a similarly detailed investigation of the broad bean *(Vicia faba)* of which the following is a summary.

Two varieties were examined in the seed and seedling stages, the green type as Green Harlington Windsor and the brown (or white) type as Giant Windsor. Differences between these two were noted in a few records but the pH ranges for most tissues were the same at most stages. All the tissues of at least two plants of each variety were examined at each stage, and three sections of each region were used for each indicator. The resulting colours were interpreted according to SMALL's R. I. M. and the pH ranges are here given in the new notation.

Seed. The cotyledon cell-contents of the soaked seeds were 56–56; the dermatogen of the radicle was 44–40, and the rest of the embryo 52–48. The integuments showed four zones, the outer layer of thickened cells being 44–40; the second layer of thickened cells and the innermost layer showed as two bands at 52–48, with the first inner parenchyma at 56–56 between the two bands at 52–48. No difference between the two varieties could be seen at this stage.

Germination. The pH of the cotyledons increased to 59–59 at the stage of unfolding of the fourth foliage leaf, and remained at that reaction in later stages. The procambial strands of both plumule and radicle increased in acidity from 52–48 to 44–40, with some 56–56 appearing in the plumule epidermis, cortex, and pith of the green variety and the 52–48 being maintained in these tissues of the brown variety, when the seedlings were grown in soil. Seedlings germinated on blotting paper developed 56–56 in ep., sep., co., mc., of plumule in both varieties. The innermost layer of the integuments changed from 52–48 to 44–40 during the early stages of germination; and during later stages acidification occurred in the first inner layer in both varieties from 56–56 to 52–48; while < 34 appeared in the outermost layer among 44–40 cells.

Stem. Sections of the stem were examined at seven stages of seedlings and also after the expansion of the sixth foliage leaf: these sections were taken—A, near the apex; B, half way down; C, just below soil-level; and D, just below the hypogeal cotyledons. Some tissues were more or less constant in their pH range throughout, such as xylem at 44–40 and cambium at 52–48 (with the exception of a few odd 56–56 cells). In the lower parts C and D 52–48 was very nearly constant at all stages in endodermis, pericycle, and phloem, as well as cambium. The 52–48 range was normal for collenchyma, and in the apical part A for ep., sep., en., pe., phl., and cambium, but 56–56 occurred in the cortex of parts A and D after the first foliage leaf expanded, with co. 59–59 in parts B and C. The 56–56 range was usual for ep., and sep. in parts B, C, D, after the first leaf expanded, and 59–59 was the usual range for the pith at all stages in all parts with some 56–56 and 52–48 records for parts below soil-level, C and D. Where slight differences between the two varieties occurred the green type was usually at 56–56 for the brown type at 59–59, or at 52–48 green and 56–56 brown type.

Root. The reaction of cell-sap in the root was more stable than in the stem. In the radicle the tissues were mainly 52–48, but even from the procambial strand stage the xylem was 44–40. In the earliest stage of germination the dermatogen and sometimes the root-cap were also 44–40. In the seedling stages the 52–48 range was constant along the length of the root in en., pe., phl., with sieve-tubes and xylem at 44–40. The piliferous layer, exodermis, and cortex were mainly 52–48 but varied to 56–56 with a few cells at 59–59. The central parenchyma varied mainly within the 59–56 range. Young lateral roots showed 59–59 in exodermis and cortex but were otherwise similar to the main root as above. The old main root also showed 59–59 in exodermis and cortex, with variations from 52–48 to 59–59 in the piliferous layer; mainly 52–48 en., pe., phl.; 44–40 xy.; and 52–48 to 56–56 in central parenchyma.

Leaf. Glandular hairs occur on the lower surface of the lamina at pH 52–48, and on the petioles at pH 44–40. The upper epidermis varied within the range 56–48, with 59–59 only in the apex of the apical leaflet of the mature plant. The lower epidermis varied similarly with 59–59 in both apex and base of the apical leaflet and also in the apex of the basal leaflet on the mature plant. The mesophyll, both palisade and spongy, was commonly 59–59 with a few variations to 56–56 and three odd records of 52–48. The xylem was 44–40 throughout, and the phloem usually 52–48 except in the less acid apical leaflets with 59–59 in most of their tissues. In the petioles the xylem and glandular hairs were 44–40, with other tissues varying mainly in the 59–56 range after the expansion of the third foliage leaf, with 52–48 common before that stage. Sections of the base of the stipule showed the tissues varying in the 56–48 range, except for xy. 44–40, with, in mature plants, the upper epidermis at 52–48 while the mesophyll, phloem, and lower epidermis were 56–56.

Flowers. The range 52–48 was common throughout the flower, and was characteristic of the primordia, glandular hairs, *young* pollen, filament, and conducting strands of anthers; also of both young and old styles, stigmas, glandular hairs, ovary wall, and all tissues of the young pod except the 44–40 conducting strands in the ovary wall, and vascular tissue in the pod. Differentiated floral components were—the pollen 52–48 when young but maturing to 56–56, the anther wall maturing from 56–56 to 59–59 with the connective at 56–56, the filament maturing from 52–48 to 52–48 and 56–56, and the young ovules at 56–56. The corolla and calyx of the two varieties differed in the greater acidity of the green type in the epidermis of the petals and sepals, all 52–48, as compared with 56–56 in the brown type, and also in the vascular strands which were 52–48 in the green type but more acid (44–40) in the brown type. The mesophyll was in the 56–48 range in both types in petals and sepals.

In the sunflower young capitulum the same 52–48 range is common, and some differentiation occurs as some tissues mature. The style acidifies to 44–40 or even <34, while the anthers become less acid from 52–48 into the 59–56 range but pollen-grains here remain 52–48; and the ovule also remains at 52–48 in all stages. The two species examined in detail differ in some vegetative tissues but are very similar in their floral tissues.

4. Seasonal Variations

(i) Leaves

Gymnosperm Leaves. DOYLE and CLINCH (1926, p. 232) using ATKINS' drop indicator method, reported a relatively constant pH for the pressed juice of conifer leaves. Their records show little variation but there are higher and lower values for the series of months given thus:—

	1925							1926
	Mai 11	June 12	July 31	Aug. 24	Sept. 25	Oct. 22	Nov. 28	Jan. 13
Juniperus	—	4.6	4.6	5.2	5.3	5.4	5.3	5.3
Cupressus	—	4.6	4.6	5.1	5.1	5.0	4.8	5.1
Larch (f. = fallen)	3.3	3.5	3.8	4.1	3.8	4.3	f.	f.
Sitka Spruce	3.5	3.4	3.5	3.7	3.7	3.7	3.7	3.7
Abies	3.7	3.5	3.7	3.5	3.8	3.7	4.3	3.7
Tsuga	3.5	3.2	3.3	3.5	3.5	3.5	3.6	3.5
Austrian Pine	3.8	3.7	3.9	3.5	3.7	3.6	3.6	3.9

The first four seem to decrease in acidity from June to August or even to October with juniper and larch, but the 0.1 or 0.2 changes in the others seem to express as high a degree of stability as occurs in some angiosperm stem tissues reported upon by REA and SMALL (1927 b), and show less variation than is shown by some tissues in angiosperm leaves examined by REA and SMALL in 1925–1926 and reported by SMALL (1929, pp. 166–182).

Angiosperm Leaves. REA and SMALL (1927 b) made a systematic examination of the tissues of a few herbs and shrubs at monthly intervals throughout an uninterrupted succession of twelve months, June 1925 to May 1926. In the case of shrubs the end leaf from a twig on the top, side and base of each shrub was taken; and from all the leaves of herbs and shrubs sections were made of the upper or apical, middle, and lower or basal part of each leaf. All the sections were examined according to the R. I. M., and the results were presented as Tables in alphabetical notation by SMALL (1929, pp. 166–182). The reproducibility of R. I. M. results is demonstrated very strongly by these Tables in which the records are sometimes the same pH range for a particular tissue in all parts of all leaves examined throughout a year, while other tissue pH ranges varied either seasonally or somewhat erratically. An extra indicator, BCG, was used in 1926 to distinguish the ranges 44–44 and 40–40.

Limnanthes douglasii. This is a striking example of uniformity within narrow limits. The petioles and bases of the leaflets were examined. The xylem was 44–40 or 44–44 in every section.

In the petioles subepidermis, cortex, endodermis and pericycle were in the pH range 59–59 in all leaves throughout the year, with the exception of one petiole in August in which all tissues other than xylem were at 56–56. The phloem, apart from the odd August petiole was also 59–59 except for December when it was 56–56 in all three petioles examined. The epidermis varied from 59–59 to 56–56, with the lower range in Sept., Nov., Dec., and Feb. to May, but 59–59 in October and January, and 52–48 in two out of three petioles in April. The ground parenchyma varied from 59–59 to 56–56, and was 52–48 in May, June, July, and January in at least two of the three petioles examined.

In the bases of leaflets the range 52–48 did not occur at all and the xylem at 44–40 or 44–44 was the only 'more acid' tissue. The range 59–59 was uniform for all tissues except in November and May when they were all 56–56, and in April when the lower epidermis and guard-cells were 56–56 with the other tissues at 59–59.

Saxifraga umbrosa. As an example of a leaf which is uniformly 'more acid' throughout the year this London Pride is as striking as the less acid *Limnanthes.* From June to December all leaf tissues were recorded as 44–40 without any exception; from January to March the range was 44–44 with the range 40–40 in the upper epidermis of the apical part of three leaves in March as the only exception. Then in April with new leaves appearing some variations were recorded: the pH range 40–40 occurred throughout the upper epidermis of the middle and basal parts of three leaves, and also in the palisade and spongy mesophyll of all parts of these same leaves mixed with the more general 44–44 cells: the other tissues of these April leaves were all 44–44. The May leaves showed more differentiation: the upper epidermis cells of all parts were mixed 44–44 and 40–40, and so were the mesophyll cells of middle and basal parts; the bundle sheath in all parts and the lower epidermis and guard-cells of middle and basal parts of these May leaves at 52–48 were the only records in a year to come above pH 4.4. The xylem and phloem were 44–40 or 44–44 throughout all leaves all the year.

With a definitely acid metabolism and the pH of cell-sap physiologically controlled by the free acids constantly present, these records are less unexpected than the rather surprising seasonal stability of the 59–56 pH range in *Limnanthes,* which does not seem to be very sensitive to carbon dioxide effects.

Rhododendron ponticum. As an example of a shrub with its leaf-cells below pH 5.2 in all tissues throughout the year this rhododendron is very satisfactory. The upper range of more acid,' 52–48, is common and is recorded as uniform all the year round in palisade and spongy mesophyll, and in bundle sheath, lower epidermis and guard-cells for all parts of all leaves; except the basal parts of three leaves in July, which were 44–40 in all tissues. The cell-sap pH in upper epidermis and hypodermis varied together from 52–48 to 44–44 and 40–40: these two layers were 52–48 in May, June and December in all parts and in November in the apical parts of three leaves: the same two layers were 44–40 from July to October in all parts and into November for the middle and basal parts of three leaves: after the December decrease of acidity, the upper epidermis and hypodermis were 44–44 alone for the apical parts of leaves from January to April, but 44–44 mixed with 40–40 for the middle and basal parts of the same leaves over the same period. The phloem in June was 52–48, like most other tissues at that time, but all the rest of the year it was at 44–40 or 44–44. The xylem was recorded as 44–40 from June to December, except for < 34 in the middle and basal parts of the 3 December leaves: from January to May xy. 44–44 occurred in all apical parts and in some middle parts of leaves: with 40–40 in the other middle parts and all basal parts: the basal parts had some xy. < 34 in January and April. The fibres of these leaves were mainly in the 44–40 range; with < 34 common among these fibres and uniform for the cuticle of the upper epidermis.

Viburnum tinus (Laurustinus). This is another shrub with a 'more acid' leaf, mainly in the 52–48 range without the lower range 44–40 in the upper epidermis and hypodermis as in rhododendron leaves. In *Laurustinus* the upper epidermal cells remained in the 52–48 range uniformly in all leaves during all the year, and so also did the lower epidermis with the exception of 44–40 in the middle and basal parts of the leaf from the top of the shrub in May. The guard-cells were also 52–48 all the time except for 56–56 in the top leaf for April. The palisade parenchyma was also 52–48 constantly except for two leaves and the apical part of the third leaf in October, and the apical parts of the top and side leaves in April. The spongy parenchyma varied much more, being 52–48 from June to August, and January to March, but 56–56 with some 52–48 in October and May, with some 59–59 in April, September and December. The xylem was recorded as 44–40 from June to December 1925 but was distinguished with BCG as 44–44 and 40–40 from January to May in 1926, the apical parts of leaves having xy. 44–44 and both ranges xy. 44–44 and xy. 40–40 occurring in lower parts of leaves. The phloem was recorded mainly as 52–48 from June 1925 to February 1926, but irregularly as 44–44 or 52–48 from March to May. Fibres varied from being seldom 52–48 to being mostly in the 44–40 range with < 34 in occasional parts of leaves. Hairs were 52–48 October to March, but within the 44–40 range April to September.

In these evergreen leaves of laurustinus the seasonal changes affected different tissues at periods of the year, but lesser acidity (59–56) appeared mainly in April and in the spongy mesophyll for both April and May.

Veronica shrub. This was a hybrid of *V. andersoni* and an unidentified species. The xylem was at 44–40 or 44–44 throughout, and the other leaf-tissues tended to be uniformly 52–48 or 56–56 at any one time, except for spongy mesophyll and bundle sheath which were 56–56 in some leaves when 52–48 was more general. The three leaves for October and the side leaf for March had all parenchyma at 56–56, but otherwise most of the tissues and all the epidermal hairs were 52–48. Apart from the four odd leaves mentioned, the upper epidermis was all 52–48; hypodermis and palisade mesophyll were each only once 56–56 instead of 52–48; the phloem, bundle sheath, lower epidermis and guard cells were all 56–56 for the top and basal leaves in May; the bundle sheath from March to July was in some leaves or parts 56–56 instead of 52–48, but the phloem and lower epidermis were 52–48 except for the five leaves mentioned. The spongy mesophyll was the most variable tissue, as might be expected in view of its carbon dioxide relations if the cell-sap is weakly buffered: the range 59–59 occurred in the basal part of the side and basal leaves for September and in the basal part of all three leaves for November and at no other time; otherwise the spongy mesophyll was nearly as often 56–56 as 52–48, except from December to February when it was all 52–48. With the one exceptional 59–59 in the spongy mesophyll, there was a distinct winter period with all parenchyma in the 52–48 range from November to February, with the top and bottom leaves for March in the same category. So that these leaves do show a seasonal stable period in winter, with lesser acidity appearing here and there during the rest of the year. In fact, the exceptions here are of more interest than the main 52–48 uniformities, which is in contrast with what occurs in the stem.

Aucuba japonica. As a contrast to the preceding acid-leaved shrubs, and to show that stability may occur in shrub-leaves as well as in the leaves of herbs, *e. g. Limnanthes,* the records for *Aucuba japonica* can be quoted. The xylem was within the 44–40 range always, except in the tips of three June leaves. The range 59–56 covers all the variation in reaction of all tissues, except for an acidification to 52–48 in all parenchyma in the side and basal leaves for June and in all three leaves for December. Most tissues were mostly at 59–59, but the guard-cells were 56–56 from February to May, and the lower epidermis was at 56–56 more often than at 59–59.

These leaves are therefore very stable in the pH of their cells, with 'more acidity' only for December, and oddly in June for two out of three leaves examined.

Dianthus caryophyllus. This has herbaceous leaves with a range of pH values for cell-sap rather similar to that of *Aucuba;* the general variation being within the 59–56 range, as 59–59 sometimes and 56–56 at other times, with general acidification of all tissues to 52–48 in one December leaf and in the basal parts of all three January leaves. The 52–48 range occurred in parenchyma only once, in one part of a leaf, outside the winter period November to February. The xylem was in the 44–40 or 44–44 range throughout, but fibres were present and showed 44–40 with <34 at most seasons. Within the winter period, apart from the four leaves mentioned above for January and December, the phloem at 52–48 in an older leaf was the only parenchyma to be 'more acid' in November; the upper and lower epidermis with its guard-cells in the middle as well as the basal parts of the three January leaves were also at 52–48; in February 52–48 extended to ue., phl., bundle sheath, le., and gc. in the middle and basal parts of all three leaves. The variation of parenchyma pH within the 59–56 range tended to be seasonal, with 56–56 commoner from February to May, and 59–59 commoner from June to January except in the upper epidermis which was mainly 56–56 with periods of 52–48 as stated above.

In SMALL's alphabetical Tables (1929) the 'c' for 56–56 is rather difficult to distinguish from the 'e' for 52–48, but these seasonal variations show up quite distinctly when the less frequent types of reaction symbols are marked with red ink. But two other kinds of leaves, primrose and privet were examined throughout the year and in these two sets the variation from tissue to tissue at the same time and from month to month was so considerable that even red-ink analysis becomes difficult.

Ligustrum vulgare. The xylem was 44–40 or 44–44 throughout; thin-walled fibres were distinguished but were never more acid than 52–48, often in the 56–56 range and noted as partly 59–59 in three separated months. This 59–59 range was the least common; it occurred in all tissues of all parts of the top leaf in January (fibres indet.), and in all tissues of the basal part of the top leaf in April, also in sp., fibres, le., and gc., in the basal part of all leaves in September, as well as in sp. and fibres of all parts of all February leaves. Otherwise the variation of pH was 56–56 or 52–48, with 52–48 more frequent from October to February and in one of the three March leaves, and 56–56 commoner in most tissues from April to August: so that in this variation within the 56–48 range a tendency towards the usual winter period of greater acidity and summer period of lesser acidity can be traced, with, however, a marked exception in the 59–59 top leaf for January, and 56–56 in spongy parenchyma for all October and November leaves, as well as the 59–59 for sp. and fibres in all February leaves.

Primula vulgaris. The variation of pH was so extensive for the leaves of this species that the mere tabulation in alphabetical symbols for each tissue in three parts of three leaves for each of twelve months (as for all the other species in this leaf series) occupies a page and a half in the original presentation of the primrose data (SMALL, 1929, pp. 170–171). Few aspects show complete stability; xylem at 44–40 or 44–44 being the only obvious one. Other tissues were relatively stable over certain periods; for example, the range 52–48 was rare from March to June which is the main live-period for ordinary primrose plants: 52–48 during this period is recorded only in the phloem of middle and basal parts of three leaves in March and three in June, also in the basal parts of three leaves in April where the fibres also were 52–48. The fibres were mostly 44–40, but records for these cells vary to 52–48, 56–56, and 59–59. The variation during this March to June period was all within the 59–56 range for ue., pp., sp., bundle sheath, le., and gc. with more 59–59 in the mesophyll during April and May than in the other tissues. In March there was only 56–56 and no 59–59; in April the tissues other than the 59–59 mesophyll were all 56–56; in May all the tissues of the apical parts of three leaves were 59–59 also the mesophyll of middle and basal parts of all leaves; in June 59—59 was confined to the mesophyll and phloem of the apical parts of three leaves. But although there was this May period of lesser acidity to a 59—59 degree, this 59—59 range is recorded for various tissues more or less throughout the later months of the year. The March to June period was preceded by two months of general acidity; January and February records are of 52—48 in all leaves and all tissues except for 56—56 in the spongy parenchyma of one January leaf.

Thus if we leave the erratic variation from July to December on one side, the variation of pH in primrose leaves show a more or less regular rise and fall of lesser acidity reaching a peak at 59—59 for all tissues in apical parts, and in the mesophyll of middle and basal parts, of all three leaves examined in May, with 59—59 in the mesophyll persisting into June only in the apical parts of all leaves. The middle and basal parts returned to 56—56 in June, and acidification to 52—48 began again in the basal parts of July leaves. Since the flower-stalk of primrose has all its tissues in the 52—40 range, it might be suggested that this local temporary mobilisation of organic acids in the flowering stem has something to do with the temporary rise in the pH or decrease of acidity of leaf cell-saps.

(ii) Stems

Rea and Small (1927 b) reported upon the pH of cells in stem tissues examined at monthly intervals throughout the same year as for leaves, June 1925 to May 1926. The same five shrubs as were used for leaves, *Aucuba*, *Ligustrum*, *Rhododendron*, *Veronica* and *Laurustinus*, were used for stems, but the herbaceous plants were all different being *Cerastium*, *Lamium*, *Senecio*, and both sun and shade plants of *Cheiranthus*. Three stems of each species were divided into upper, middle, and lower parts each month; and a sufficient number of sections were cut from each part to allow immersion of two or three sections in each of 5 or 6 indicators; and the colours given by 9 or 10 different parts, tissues and hairs, were noted systematically by Rea. The resulting colour schedules were translated into pH ranges by Small, and presented as alphabetical symbols in a series of Tables which are now, over 25 years later, being re-analysed by Small with perhaps an improved understanding and certainly with an increasing admiration for the patient hard work of his one-time research assistant, Miss M. W. Rea. As for the leaves, bromo-cresol green (BCG) was used from January 1926, and gave the distinction of the xylem as 44—44 or 40—40, or both, instead of merely 44—40.

Cerastium tomentosum. This has mostly less acid (59—56) vegetative and flowering stems. The Vs. was examined throughout the year and, with very few exceptions showed 59—59 or 56—56 in ep., sep., co., phl., mr., and mc.: the exceptional appearances of 52—48 in these tissues were confined to ep. and phl. of middle and lower parts of January and July stems, with sep. and mr. 52—48 in some parts of the July stems. The range 52—48 also occurred for hairs from November to July, with 56—48 in August and 56—56 in September. The pH of ep., sep., and phl. fluctuated from 59–59 to 56–56 but the cortex was always 59–59, with some walls 44–40, except for 56–56 in all February stems and parts of the March stems. The endodermis was 59–59 from May to September but showed 52–48 and 56–56 sporadically from October to April with walls occasionally noted as 44–40 or 44–44. The medullary tissues mr., mc., varied together from 59–59 to 56–56 in odd months, with 52–48 for mr. in parts of July and April stems. The xylem was 44–40 or 44–44 throughout. The most complex records are those for the pericycle cell-contents and cell-walls, the contents varying from 59–59, 56–56, to 52–48 inside cell-walls at 44–40 or 44–44 or < 34. The regularities for other tissues leave no doubt that this diversity of pH in the pericycle is a real phenomenon, but it does not seem to be seasonal, even for part of the year, in the way that has been traced above for certain tissues of primrose leaves.

Cheiranthus cheiri. The stems of these wallflower plants, both in sun and in shade, yielded records of a nearly uniform pH 59–59 in all tissues except the 44–40 or 44–44 xylem.

In the sun plants the exceptions were as follows: in December all three stems showed "more acid," being 52–48 for en., and 44–40 for phl., mr., mc.; in January all three stems were partly 52–48 in ep. and en.; in October one stem was 56–56 in ep., sep., co., en., while the other two October stems were partly 56–56 in ep., en., and mr.; apart from collenchymatous cell walls at 52–48 noted for July and November in the cortex, these are the only variations from 59–59 for all tissues and for epidermal hairs.

In the shade the uniformity of 59–59 was even greater. The subepidermis, cortex, pericycle, phloem and central medulla remained at 59–59 throughout the year without any exception. The epidermis was 59–59 in September and from November to May, with 56–56 in the lower parts of all stems in June and July, and 56–56 noted for cell-walls with 59–59 cell contents in October, and 44–40 for cuticle in the lower parts of August stems. The endodermis was 59–59 from March to October, with 52–48 for en. in the lower parts of 3 November stems and in the lower parts of one stem out of three for each of the months December, January, March, showing a slight winter period of "more acid" pH affecting only one tissue, but the December acidity of the lower part of the stem affected some cells in the medullary rays of lower parts of all three stems; these cells were at 44–40 and this range in the mr. tissue was also noted for the lower parts of all three stems for June, July, August and of two stems in September. The hairs were 59–59 in November-December, and 56–48 June to September, but were not recorded for January to May.

The fact, that about twenty-five thousand observations can be made upon 216 stems from 72 plants of one species, half from a sunny position in one garden and half from a shaded position in another garden, without yielding more than a few variations in the pH range for any one tissue, bears witness to both the stability of reaction of the cell saps in these plants and the reliability of SMALL's Range Indicator Method. It has been suggested that uniformity of records may be due to diffusion of cell-sap as an alcohol effect, but the differentiation obtained for certain tissues if not for others, such as in the pericycle of *Cerastium*, demonstrates that the uniformities are as natural and real as the diversities.

Lamium purpureum. In the stems of the deadnettle, as if to answer the diffusion suggestion, some of the uniformities are in time rather than across sections with some tissues uniformly differentiated from others which vary in other pH ranges. Collenchyma and the epidermal hairs were at 52–48 and xylem at 44–40 or 44–44 throughout all stems always. The epidermis was self-coloured pink. The range 52–48 appeared, apart from coll. and hairs, only once, in the subepidermis of the 3 June stems. Otherwise all the variation was from 59–59 to 56–56 with 59–56 only in mc. of the middle and lower parts of 3 July stems. Apart from those 59–56 parts in July, the central medulla was 59–59 from May to October and from December to February, being 56–56 in all 3 November stems, 2 April stems, the upper parts of March stems, and the middle and lower parts of the third April stem. The variation from 59–59 to 56–56 in other tissues was for shorter periods and also from one part to another of the same stem, so that no clearly seasonal trends can be seen for any one of these tissues; for example, sep., co., en., phl., mr., mc. were all 56–56 in 3 November stems, and all 59–59 in 3 December stems, but in 2 January stems sep., co., en., phl. were 56–56 while mr., mc. were 59–59, and the third January stem had en. 56–56 with sep., co., phl., mr., mc. at 59–59. The cortex and endodermis tended to vary together and were both at 59–59 in all stems for July and August but not for any other two consecutive months.

Senecio vulgaris. Like the other three herbaceous stems examined, the groundsel stem has most of its parenchyma in the pH range 59–56 (mostly 56–56) throughout the year, except for two "more acid" tissues, xylem at 44–40 or 44–44, and collenchyma at 52–48 always. The range 52–48 occurred in the epidermis of middle and lower parts of 3 May stems, and in the hairs of all stems in January and February. In all other tissues the variations were from 56–56 to 59–59, with some seasonal periods. From June to November for all stems ep., sep., pe., phl. were at pH 56–56 with the one exception of some 59–59 in the lower parts of the 3 July stems. During this June-November period of 56–56 for four outer tissues the inner tissues, mr. and mc., were at 59–59 for all stems in June to August and in October, and at 56–56 in November; in all September stems mr. was 56–56 but mc. varied with the part for two stems and was 56–56 in the third. In January and March, ep., sep., co., en., pe., phl., mr., mc. were all at 56–56, but in February, April, and May the uniformity did not occur, various combinations of 56–56 and 59–59 giving variety within the 59–56 range. The main long uniform periods were:— all year for xylem and collenchyma as above; March to December 56–56 for hairs; January to April, and June to November 56–56 for pericycle and phloem (with 59–59 in December and May); June to November 56–56 for epidermis and subepidermis. The cortex and endodermis tended to vary together from month to month and part to part of the same stems, no two successive months showing all parts the same except 59–59 for co., en., in all parts of all stems for April and May.

Aucuba japonica. The seasonal variation of cell-saps in the stems of *Aucuba* were very similar to those found in the leaf. The range 59–59 was most common, but for top and side positions in May, for the top in June, and all three positions in October most tissues were 56–56 in both stems and leaves; while for side and base in June and all positions in December most tissues were 52–48 in both stems and leaves. There was also some development of 52–48 in most tissues of the basal stems for April and May, which was not paralleled in the leaves at that period. Outside these periods of general "more acid" values, 52–48 occurred only in stem epidermis where it was frequent in some or all stems except for a run of 56–56 from August to October. Each tissue varied from 59–59 to 56–56, except for the more acid cases cited above; with 59–59 general in August, September, November, and January to April, and with 56–56 general in 2 May stems, 1 June stem, and all 3 October stems.

Rhododendron ponticum. As in the leaf of this "more acid" shrub there was no record of a pH higher than 5.2 in any cells at any time of year. The stem epidermis was normally 52–48 like the lower epidermis of the leaves, and not 44–40 like the upper epidermis and hypodermis of the leaf, except for 44–44 in the middle and lower parts of all 3 stems for April and in all parts of all 3 stems in February. The same 52–48 range was constant throughout all stems all the year in sep., co., en., mr., and mc., the only exception being 44–44 for sep. in the 3 February stems. The varying tissues were pericycle, phloem, and xylem. The xylem was 44–40 or 44–44, but mainly 40–40 with <34 frequent in addition to 44–40 or 40–40. The phloem varied, showing 52–48 in June, September, December, May, and in one April stem, recorded as 44–40 in July, August, October, November, and as 44–44 January, February, March, and in 2 April stems. The pericycle showed some 52–48 in June, July, September, October, and some 44–40 from June to December, also 40–40 in January, February, March, and in the middle and lower parts of stems for April and May, with 44–44 in the upper parts of the April and May stems: the pericycle cell walls were frequently at <34 for most of the year. The pH ranges for the young stems of rhododendron are thus very similar to those of the leaves, and are so constant for any one stem tissue that seasonal variations have not been traced, not even the May-June or November-December decreases of acidity such as occurred in the leaves.

Viburnum tinus. The stems, like the leaves, of Laurustinus, have only a few records of pH values above pH 5.2. In the leaves there were about 6 cases of 59–59 only in the spongy mesophyll, and about 15 cases of 56–56 of which all but six were also in the same tissue: only one April stem showed any tendency to be less acid. The basal stem for September was even more definitely "less acid," being 59–59 in all parts for ep., sep., co., en., phl., and in upper and middle parts in pe., 59–59 and 56–56 in mr. and mc., also 56–56 for hairs, while the range 59–59 occurred nowhere else in a page and a half of tabulated variations. The range 56–56 occurred in parts of three July stems in sep.. co., en., also in two April stems in co. and en., as well as in mr., mc., and hairs in the above-mentioned September stem. Some of the tissues were more or less constant in pH, but pericycle, phloem and hairs were variable. Apart from the odd September stem — the ep., sep., co., en., mr. and mc. were almost constantly at 52–48, with 44–44 for ep. in one May stem; and the above-mentioned 56–56 variations. The xylem varied within the range 44–40 with parts frequently < 34, and the pericycle of the middle and lower parts of stems varied mostly over these same ranges with < 34 frequent, but in the upper parts of stems the pericycle was at 52–48 except in September and March (indet. in January, February). The phloem was usually 52–48 but 44–44 occurred in some stem parts in March, April, May, and 44–40 in the lower parts of 2 June stems. The cuticle was almost uniformly 40–40 where it was noted, but < 34 for all March and December stems. The pH ranges recorded for epidermal hairs show a curious monthly sequence with little variation within any one month, as if the cell-contents of these hairs were subject to some seasonal influence: this hair pH was 40–40 from January to April without exception, a mixture of 52–48 and 44–44 in May, 44–40 with < 34 in June, 52–48 in July, 44–40 with < 34 in August, < 34 for 2 stems in September with 56–56 in the odd September stem, 52–48 in October, and 44–40 throughout November and December before beginning the long 40–40 period January to April.

Ligustrum vulgare. In the leaves of privet the range 59–59 was rare, being confined to the top leaf for January, the basal part of all leaves in September and of one leaf in April, and the sp. and fibres of all February leaves. The distribution of this 59–59 range in privet stems is quite different: 59–59 is not present in the records for March, May, June, July, and December; but it is present for one August stem, and all stems in September, October, November, January, February and for one stem in April. There is, therefore, a strong autumn-winter period for 59–59 in privet stems, with small extensions for an August beginning and an April ending. This lesser acidity started with one August stem of which the upper part had all its tissues at 59–59 (except the xylem), the middle part with all its tissues at 59–59 (except xy., pe.), the lower part with 59–59 only in its cortex and endodermis. The range 59–59 occurred again in the epidermis of the upper part of 1 September stem, and in the sub-epidermis of 2 September stems. The range 59–59 is recorded for cortex and endodermis in all stems from the one in August to all 3 stems for September, October, November, January, and 1 each for February, and April. This 59–59 lesser acidity extended to pe., phl., mr. and mc. in the upper parts of stems in the 1 stem in August, and all 3 stems for September and October; 59–59 also extended to phloem in November (mid. and lowest parts) and in April (lowest parts of 2 stems), and to mr. in 2 stems as well as phl. in all 3 January stems; also to mr. and mc. in April, August, September (mid. part of 1 stem each), and in February (upper part of 3 stems).

The range 52–48 occurred in all hairs, and in some tissues in every month, being commoner as greater acidity in ep. and sep. than 56–56 from October to May; from June to September these two layers are mostly 56–56 or 59–59. The cortex and endodermis tend to be 56–56 outside their 59–59 period, but are at 52–48 in some parts of some stems in March, April, July, and December. The medullary tissues on the other hand tend to be more 52–48 with some 59–59 in August to February, and more 56–56 in May, June, and August.

In fact, the seasonal history of pH in the cells of privet stems could be seen in the data to have such precision and regularity that SMALL tried to draw a diagram of it all by filling up the thousand and two spaces for which records were available, 9 parts of stems for each of 12 months for 9 tissues, with 30 records for hairs. Each of these 1002 records of a given pH range involved observations on all the tissues of 2 or 3 sections immersed in each of 5 or 6 indicators, 5 for 7 months and 6 for 5 months giving an "indicator × month" factor of × 65, and a section factor of × 2.5, so that the total observations amount to 81 × 65 × 2.5 or 13,162 for this one-page diagram of privet stem pH ranges (Fig. 1).

Veronica sp. × *V. andersoni.* The leaves of this shrub had such a wide and constant development of the pH range 52–48, that the exceptions seemed to be of more interest than the uniformities. The stem also had 52–48 common, and all hairs were constantly 52–48, but there was a large development of 56–56 in the upper and middle parts of stems, with 52–48 so often in the lower parts of the same stems that the lower parts were tabulated separately. In these lower parts of stems the range 52–48 was so general that it is simpler to note the less common cases of 56–56 and the rare 59–59 records. The 59–59 records are for mc. in 3 September stems, 2 October stems, and 1 November stem (lowest parts), also for sep. and en. in 2 and co. in all 3 of the September stems. These same 3 stems in their upper and middle parts had 59–59 in co., and in their upper parts for en. and mc. In the lowest parts 56–56 occurred in all the other tissues of the September stems, and in most tissues of:— all 3 July and October stems, one in November, 1 in March and 1 in April: otherwise 56–56 was rather sporadic in its appearances; such as in co. for 2 stems in August (its only record for that month); in mr. for 1 November stem; in co., mc. for 1 in January; in ep. for 1 and co., en., for another 1 in March; and co., en., mr., mc. for 1 in May. In their lower parts, 9 out of 36 stems were mostly 56–56, and the other 27 were mainly 52–48.

The distribution of pH ranges in the upper and middle parts of these 36 stems was rather different: 16 out of 72 of these stem parts were mostly 52–48 and the other 56 were mainly 56–56. Most of the tissues in any one part of a stem were at the same pH, either 56–56 or 52–48, except for the 44–40 or 44–44 xylem and the uniformly 52–48 hairs: but the cortex was in the 59–56 range in the upper and middle stem parts except for a winter

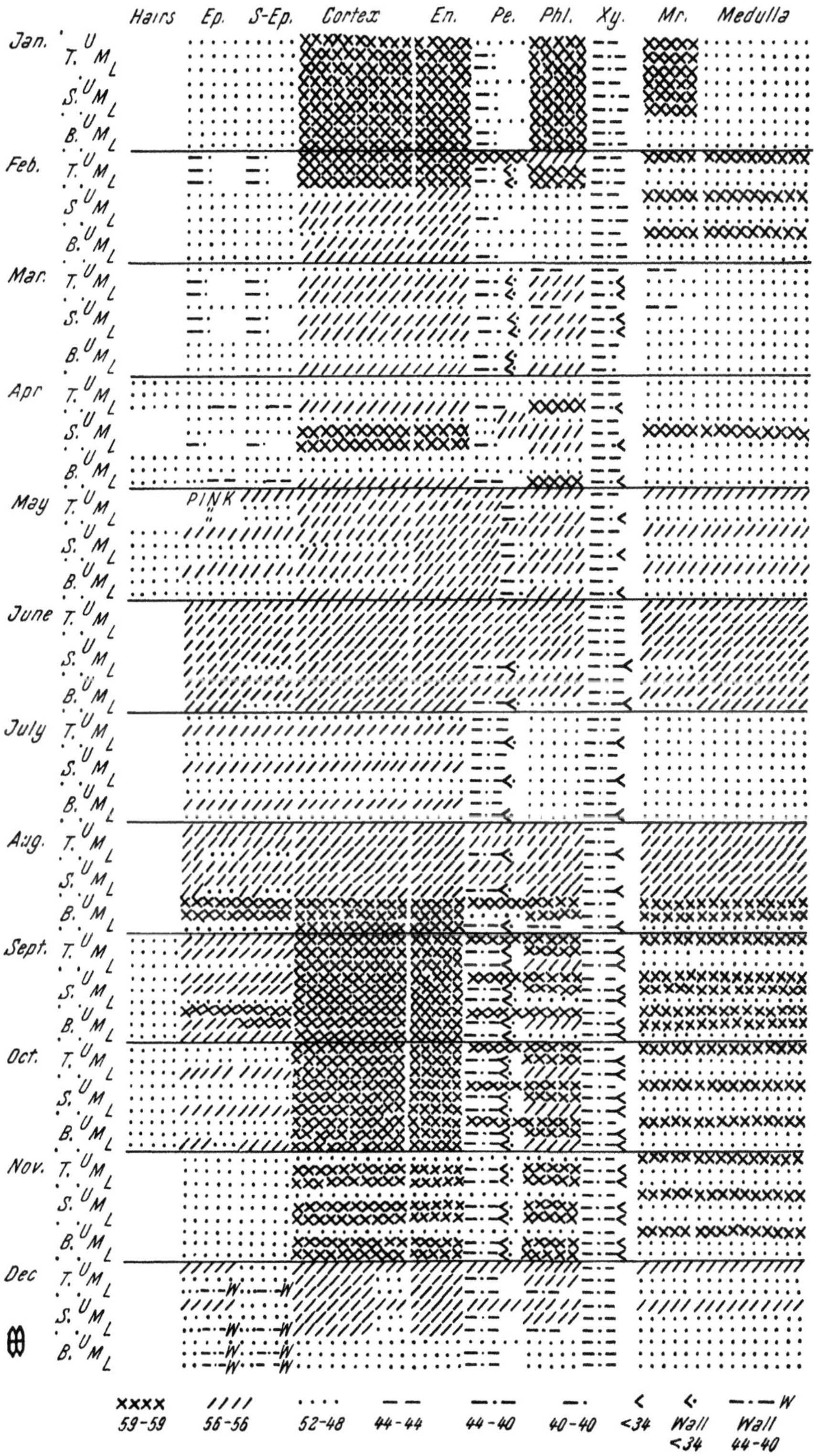

Fig. 1. Summary of seasonal records of cell pH ranges for stems of privet (*Ligustrum vulgare*).

period indicated by 52–48 in 2 November, 3 December, 2 January, 3 February, 1 March, and 1 May stems, also 1 in August. Similarly the central medulla was 52–48 only in 1 November, 3 December, 2 January, 2 May, and 3 August stems, which is not as strongly seasonal as the distribution for cortex pH.

(iii) Summary of seasonal variations of pH in Stems and Leaves

Shrubs — leaves and stems. Five shrubs were examined in detail, both stems and leaves, at monthly intervals throughout one year. *Rhododendron ponticum* was at 52–48 in most of its parenchymatous tissues and 44–40 in pe., phl., xy., and fibres in the leaves. The main variation occurred in the upper epidermis and hypodermis of the leaves which were usually 44–40 or 44–40, but showed 52–48 in May, June, and also in December: in the stem ep. and sep. were normally 52–48 in most months with 44–44 in February. *Viburnum tinus* was similarly 52–48 in most parenchyma in most months, but 59–59 was shown in the spongy mesophyll of the leaf in December, September and April, and also in the basal stem for September for nearly all tissues of all parts, but one month is hardly a season. The *Veronica* shrub showed a marked difference between the lower "more acid" stem parts and the upper and middle parts. The leaves and lower stem parts were mainly 52–48, with 56–56 in some stems but dominant only in the 3 October stems and in 1 March stem; in contrast the upper and middle parts of stems were mainly 56–56 with 52–48 occurring in a vaguely seasonal way, November to March, but also in May and August. This seasonal variation is the only one in these three acidic shrubs which extended over several consecutive months.

The other two shrubs were *Aucuba* and *Ligustrum*. *Aucuba japonica* maintained the low acidity of 59–59 in both stems and leaves in most tissues most of the time, so that any variations were towards increased acidity: the leaves showed 52–48 only in December and June, but 56–56 was general in May and in one June leaf so that greater acidity seems to have occurred in a short summer season as well as in December. This happened also in the stems, with 52–48 in one May and one June stem and 56–56 dominant in the other two in both months: the stem also was 52–48 in December, so that stems and leaves tended to vary then in the same way. The *Veronica* more acid period November to May has no counterpart in *Aucuba*.

In *Ligustrum* leaf (Fig. 1) there was a general 59–59 in one leaf in January, in one part of one leaf in April, and in one part of three leaves in September, with 52–48 in a winter period October to December and in two January leaves. This winter 52–48 occurred in epidermis and subepidermis of the stem, November to April, with the central medulla rather similar; but there was the contrasting period of 59–59 in the cortex, endodermis, and phloem, with slight extensions to some other tissues, August to February, and with relatively more acidity again in December. This seasonal variation in opposite directions in two sets of tissues in the privet stem rather upsets all conclusions drawn from variations in mixed expressed saps since the average pH of these pressed juices might remain similar with tissue pH ranges actually varying from 56–56 to 52–48 as in ep., sep. and pith, and from 52–48 or 56–56 to 59–59 as in co., en., phl.

Herbs — leaves. The four kinds of leaves from herbs showed much the same range of variation as the stems and leaves of shrubs. *Saxifraga umbrosa* leaf showed very little variation all through the year, always in the 44–40 range for all tissues, except a little 52–48 in some cells of the lower side of all leaves in May. *Limnanthes douglasii* leaflet bases and petioles were just as regularly and continuously at 59–59 at all times in all tissues, with some 56–56 in epidermis at odd times, in all leaflet bases for November and in one petiole for August. No really seasonal variations were found in either the more acid or the less acid of these two sets. *Dianthus caryophyllus* leaves showed much variation in all their tissues. Variations within the range 59–56, from 59–59 to 56–56, were common and rather erratic in most tissues, different parts of the same leaf or different leaves in the same months showing 59–59 or 56–56; but 52–48 was general only in one December leaf and in the lower parts of all 3 January leaves, so that here again there was a December increase of acidity (as in privet stem) which in *Dianthus* leaves extended in one part to January, and also into February in ue., phl., le., gc., and bundle sheath of middle and lower parts of all leaves. Outside these three winter months 52–48 occurred rarely in only one or two tissues in odd leaves, so that here there was a definite seasonal acidification in December to February.

The leaves of *Primula vulgaris* showed even more variation which has already been described in some detail, so that here it is necessary only to emphasise the spring season of growth and flowering of the primrose and the concomitant rise of pH from a general 52–48 January-February, to 56–56 March-April, and to 59–59 May-June, without giving the same kind of significance to the 56–56 with 52–48 in July and August. The seasonal increase in pH values in the first six months of the year is here quite obvious in all parts and especially in the apical parts of primrose leaves.

Herbs — stems. All the herbaceous stems showed some development of 59–59 which was more or less dominant in three of the four species examined but only occasional in *Senecio.* In the groundsel, *Senecio vulgaris,* the common pH range was 56–56, but 59–59 occurred in most parenchyma, including mr. and mc. of all 3 December stems, in these medullary tissues over a summer seasonal period, April to July, and also in October; so that 59–59 was not confined to the summer even in mr. and mc., but in October co., en. varied in parts but were mostly 56–56, whereas in April, May, July, and December co. and en. were usually 59–59 as for mr. and mc. The summer period of lower acidity affected all the bulky tissues, but the difference was only that between 59–59 and 56–56. The range 52–48 was more or less confined to collenchyma and the hairs in January-February.

Lamium purpureum (deadnettle) stems had a natural pink colour in epidermis; the pericycle seldom yielded a definite pH range, and the other parenchymatous tissues tended to vary so erratically from one part to another of the same stems that few definite seasonal variations can be traced in any one set or even in any one tissue, except perhaps a dominance of the lesser acidity 59–59 in mr. and mc. from May to August; also for co. and en. in May, July, August with 56–56 for co. and en. in all June stems. The range 52–48 is almost confined to collenchyma and hairs, so that there is no "more acid" winter period in deadnettle stems, although there are indications of a summer period in certain tissues of higher pH, pH 5.9 rather than pH 5.6.

Cerastium tomentosum (hairy chickweed) stems showed a similar dominance of the 59–56 pH range. Apart from the 52–48 hairs, this 52–48 pH range occurred sporadically and only rarely except in the middle and lower parts of stems in January and July. The 56–56 range occurred for most parenchymatous cells only in February for all parts of 3 stems, but for the upper and middle parts of all 3 stems in November, March, and for some tissues in April, so that with 59–59 most common there is a slight spring drop to 56–56 in February, March, April, but no sign of a more acidic winter period.

Cheiranthus cheiri. The other two sets for herbaceous stems consisted of a comparison throughout the year of the stem tissues of wallflower plants growing (a) in a sunny position and (b) in a shady position, in different gardens. Nearly all parenchyma in both sets of stems was at 59–59 throughout the year with a few drops to 56–56 and still fewer to 52–48, or even to 44–40 in phl., mr. and mc. of all 3 December "sun"-stems. One October "sun"-stem had 56–56 in ep., sep., co., en., of all its parts, but this was only one out of 72 stems. At other times of the year only a rare and usually only partial development of 56–56 in the epidermis made slight breaks in the monotony of 59–59 in ep., sep., co., en., pe., phl., mc., with some 44–40 in the medullary rays in both sun and shade stems from June to December. There was, therefore, no seasonal variation of cell-sap pH in either sun or shade conditions for wallflower stems.

Summing this review of our year-long surveys of pH in leaves and stems, seasonal variations which were more or less definite often had December as an exceptional month in otherwise continuous series or even when there was a distinguishable summer series isolated from a somewhat similar December. Rhododendron leaves showed this in their May-June and December decreases of acidity (44–40 to 52–48); *Veronica* stems in their upper and middle parts had increased acidity (56–56 to 52–48) in a winter period, November to March; *Viburnum* had no seasonal variation. *Aucuba* in both leaves and stems had a slight increase of acidity (59–59 to 56–56) in May and partly in June, with 52–48 in some other stems and leaves in June, but this 52–48 appeared also in December. These were increases of acidity, instead of the May-June and December decreases as in *Rhododendron. Ligustrum* leaves showed increased acidity (59 or 56 to 52–48) in a winter period October-January, and also a 52–48 winter period November-April in ep., sep., and mc. of stems; but in these stems there was a complementary re-distribution of pH in co., en., and phl. up to 59–59 when ep., sep., and mc. were down at 52–48, with an equalisation of all or most tissues at 56–56 in May, June, and August. This sequence was broken in July when nearly all tissues of upper and lower parts of stems were "more acid" at 52–48, with ep., sep., co., en. in the middle parts at 56–56 (see Fig. 1). This striking seasonal re-distribution stands out as something very definite which was missed in the 1929 analysis.

Among the herbaceous stems and leaves, there was in *Senecio* a December decrease of acidity (56–56 to 59–59) in mr. and mc., which decrease also characterised these same medullary tissues from April to July, as a summer period of decreased acidity for these tissues, and also for co., and en., so that this summer period of lesser acidity (59–59) was general for the bulky tissues of groundsel stems. A similar 59–59 summer period can be traced for *Lamium* in May to August for mr., mc., co., en., with 56–56 in for co., en., in June as a break in that sequence. In *Cerastium* the only sequence appears to be a slight spring acidification (59–59 to 56–56) in February to April. In all the other material examined, whether it was dominantly within the "more acid" 52–40 range or within the "less acid" 59–56 range no very definite seasonal variations could be traced, even within these ranges.

Re-calibration of all indicators

It is with all these thousands of estimations of cell-sap pH ranges as a background that Small is now reviewing the other work which has been done, both before and after 1925/26. The recent adoption of a value of 0.2443 V for the E_0 of the saturated calomel electrode at 25° C. (Marconi Instruments Ltd., 1951) may result in the re-calibration of all indicators. According to this booklet (*op. cit.* p. 44) this value for E_0 is about "2 mV below that hitherto accepted and gives rise to pH values roughly 0.03–0.04 units above values calculated on older scales." The main standard for the new pH scale is quoted as being at pH 4.005 at 25° C. on the new scale, whereas Sörensen gave pH 3.97 for 0.05 M potassium hydrogen phthalate. If all indicators are to be in future re-calibrated to match the new empirical pH scale, the values given for all these R. I. M. pH ranges may require some revision. At present they are all consistent within themselves as a comparative calibrated system, and if .03–.04 is to be added systematically to all pre-1950 indicator pH values in future, these R. I. M. pH *numbers* may have 0.3 or 0.4 added to all the simple 59, 56, 52, 48, 44 and 40, figures which have here been used to indicate pH-ranges. But it seems to the present writer that this would be an unnecessary complication for general survey work, although in cases where critical comparisons are being made, between pH values estimated with a glass electrode with an accuracy of $\pm$ 0.02 and corresponding pH values estimated with indicators to the nearest 0.1 or 0.2, the glass electrode will always have an advantage over indicators where the material is available in free solution in sufficient quantity. The last qualification is the operative factor in dealing with normal sized cells arranged in tissues which have been demonstrated as differentiated in their pH values in many plant stems and leaves. A preliminary survey is necessary for any research material from which it is desired to express cell-juices. If it is as uniform as *Limnanthes* leaves and stems, mixing cell-saps might be justified as mixing saps at a very similar pH value, but with most stems or leaves such a survey would show that such mixing would be quite unjustifiable as part of a procedure for estimating the pH value of plant cell-saps.

5. Maturing changes and gradients

The seasonal variations of cell-sap, as given in the preceding section are empirically observed data. Explanations must apply to individual cases, because of the variability of observed increases or decreases of acidity in various tissues at various seasons. With all the possible variables it is easy to theorise in the absence of restricting facts; there may be fluctuations or seasonal changes in (1) acid-producing processes, or (2) acid-using processes, or (3) buffer-producing or buffer-using processes. As a speculation it might be suggested that the winter decreases of acidity in cells towards the outside of stems take place in cells with an acid-producing metabolism which is governed by temperature and/or light conditions; while the summer periods of lesser acidity for inner tissues take place in cells where increased efficiency of metabolism results in the more rapid utilisation of acids and/or an increase in the buffer index of the sap in the region of the natural pH. In particular, the phosphate buffer-index for tissues maintained at 59–59 may be suggested as a possible stabilising factor in these cases.

In fact, it appears to the writer that many of the records of pH values (especially between pH 5 and pH 7) for plant juices are to be correlated with metabolic changes in the buffer-systems or buffer-complexes rather than with increased or decreased production of acids (other than carbon dioxide). Given an inactivation of part of the buffer-complex, by utilisation in the cytoplasm or otherwise, the pH would change without the production of greater or lesser quantities of actual acids, the degree of dissociation of which would vary with the proportions of the constituents of the buffer systems (see RICCI, 1952, p. 196). A good example of relatively stable pH value maintained during 21 weeks of growth and maturing in spite of a doubling in the amounts of titratable acids was reported by HURD-KARRER (1928). The titratable acids of both Jenkin and White Odessa Wheat juices increased almost regularly in from 12 to 21 weeks of growth so that the semi-final ratios were 11.6/6.2 and 14.7/7.8, while the pH varied in the same period around $5.7 \pm .05$ or 5.7–5.8 using only a first decimal. HURD-KARRER *(op. cit.)* quite logically attributed this maintenance of pH within narrow limits to the efficiency of the buffer-systems present in wheat plants.

There are, in the R. I. M. surveys, many cases where certain tissues, such as xylem and pericyle, increase in acidity from top to lower parts of stems while other tissues, such as cortex and pith, decrease in acidity with the top at 52–48 and either one or both of the middle and lower parts at 56–56 or sometimes 59–59. This gives gradients in opposite directions for different sets of tissues, which are known from the sunflower and broad bean data to be *maturing* changes. SIDERIS and YOUNG (1944) found a gradient in pineapple leaves but none in the stems: expressed sap from leaf-parts, base to apex, gave pH values of 5.5–4.2–3.5–3.3. These gradients were correlated with similar series for content of citric acid, ascorbic acid, chlorophyll, and carotenoid content; and it was noted that although there was some measure of agreement of pH with total acidity "the great amounts of buffer substances in certain sections made perfect agreement between titratable acidity and pH values impossible."

Maturing changes which are definitely not just structural gradients have been followed in fruits. BARTHOLOMEW (1923) reported a good series for lemon at different stages of maturity, using the hydrogen electrode and second decimal figures, and distinguishing in the second decimal between stylar and stem ends of the fruits, thus:—

Dates 1920–21	Sept.	Oct.	Nov.	Dec.	Jan.	Feb.	Mar.	Apr.	May	June	July
Stylar end	4.46	2.91	2.64	2.54	2.60	2.57	2.33	2.30	2.23	2.29	2.29
Stem end	4.46	3.08	2.71	2.50	2.64	2.54	2.36	2.33	2.27	2.33	2.33

BARTHOLOMEW noted that, while total acidity of lemon juice increases with age, the pH changed very little after the fruit was about 3.8 cm. in diameter; this is a natural result of the increase in buffer index value with increased citric acid concentration: apart from second decimal aleatory variations, the pH of the juice in these records reached a modern value of 2.3 in March and did not subsequently show any significant change. GUSTAFSON (1927) made a similar study of ripening tomatoes, and found that the pH of the juice fell to the 5th week, approximately thus:— 4.9–4.9–4.8–4.2, and then rose slightly to 4.3 in the 7th and 4.35 in the 8th week. He noted that low pH values are associated with the period of rapid growth, a correlation which on modern views might be part of a chain of events involving greater efficiency of a given concentration of growth hormone at these 4.2–4.35 pH values. This kind of gradient, with a more acid apical zone, is revealed in privet stems for the medulla in September to January (Fig. 1) but might apply to cortex and endodermis in February, March and July.

Another study of pH in relation to maturing of fruits is that on apples by KROTKOV, WILSON and STREET (1951). Although this is still another series of pH values for juices and not for individual cells, apples like lemons have reasonably uniform juicy tissues. The juice was obtained by freezing at —18° C., either whole young apples or peeled-quartered-cored larger and older apples; after freezing, all apples were chopped and pulped, then the juice was obtained by filtration in a Bucher funnel by vacuum suction. Their conclusion from work in 1942–1943 on apples from three different sources was that ontogenetic changes were remarkably uniform in the apple fruits used. Later results in 1948–1949 were from apples all from one tree. Respiratory changes, and variations in sugars and organic acids, mainly malic, as well as pH values, were followed from June to September in fruits direct from the tree, and afterwards in storage. The pH of the juice fell rapidly in June, as measured at weekly intervals, from about pH 5.7 at fruit-setting to about pH 3.1 at the end of June, to pH 2.8 in July, and back to 3.1 until all the apples were picked in September; after which the pH rose slowly to about pH 3.5 during storage until March. These values are for 1942–1943; for 1948–1949 the initial June value was about pH 4.8, the lowest value (July) about pH 2.8, rising to 3.1 in September, and to about 3.7 during storage until May. The increase of pH, after the first six weeks of acidification, was so regular that the pH value "is a better indicator of the ontogenetic stage of an apple fruit, than its respiratory rate and either carbohydrate or acid content."

An interesting example of the application of the R. I. M., combined with examination of expressed juice, is the comparison of masculine and feminine plants of *Lychnis dioica* by STANFIELD (1937), together with some data on variations of pH during anthesis.

In the actively blooming stage of the plant the stem tissues of staminate plants tended towards higher pH values, thus:—

staminate — cortex 4.8–6.0, pith 5.0–6.0;
pistillate — cortex 4.8–5.2, pith 4.8–5.2.

As cortex and pith yield most of the juice, it is noteworthy that the expressed juices of similar plants varied in the opposite sense, thus—staminate plants during anthesis yielded juice of pH 6.0 early, 5.5 mid, 5.0 late, and 6.8–5.1 later; while pistillate plants during anthesis yielded juice of pH 6.4 early, 6.6 mid, 6.3 late, 6.7–5.9 later: both start at the pre-blooming rosette stage with pH 6.8, so that staminate juice becomes more acid, while pistillate juice stays more or less constant. This looks very like a carbon dioxide effect with the buffers at two different concentrations or compositions. STANFIELD (1944) reports more phosphate in staminate plants, but there may also be organic acids such as citrates.

The main floral tissues were reported as pH 4.8–5.2 for ovary-wall and placenta, with pH 5.0–6.0 for the male pseudo-receptacle so that the male floral tissue is less acid, as in the main stem tissues. More calcium oxalate crystals occur in the placenta and ovary wall than in the pseudo-receptacle; therefore there is more oxalic acid with greater acidity. STANFIELD (1944) again reports initial pH values of 6.8 for pressed sap from *tops* of both sexes, with staminate 6.0 and pistillate 6.4 in the early flowering stage; also for *root saps* 5.8–6.0 later for staminate and 6.0–6.2 later for pistillate plants.

Another floral observation was made by RIJVEN (1952) who found, using SMALL's R. I. M., that the ovular sap of *Capsella bursa-pastoris* remained about pH 5.9 or pH 6 while the embryo increased from 100 μ to 700 μ in length, and that growth of the embryo was better at pH 6.6 than at or below pH 5.4.

ROBERTS and DOYLE (1938, p. 662) reported a general increase of juice pH values with age, in a comparison of young and old conifer leaves. This age-difference of 0.2 to 1.4 pH units varied with the season which affected the young leaves rather more than the old leaves, as is shown in the following selected results:

		June	July	Oct.-Nov.
Picea breweriana	young	3.4	3.9	3.1
	old	4.8	4.8	5.3
Pinus excelsa	young	3.7	3.9	3.9
	old	4.4	4.4	4.5
Taxus baccata	young	4.7	4.9	5.3
	old	5.3	5.4	5.5

6. Diurnal Variations

(i) Diurnal Variations in non-succulents

ROHDE (1917) reported for *Spirogyra* sap pH 6.7 during the night and pH 5.9—5.5 during the day, and noted a similar acidification in light for the epidermis of carnation and tulip flowers; but TRUOG (1919) found juices of clover, alfalfa, soybean, and buckwheat plants more acid in the morning than in the afternoon. CLEVENGER (1919) found that the pH of sap from stems and leaves of cow-pea rose in the afternoon, and fell during the night reaching a minimum in the morning, thus agreeing with TRUOG's results. HAAS (1920) on the other hand records a higher pH value for juice of corn seedlings in decreased illumination and HURD (1923) found an irregular variation in morning and afternoon saps of maize, mostly in the second decimal figures, but she reported later (HURD, 1924) on wheat sap pH as being lower in the morning and higher in the afternoon, without any corresponding variation in titratable acidity. Total acidity is often higher in the morning than later in the day (see CZAPEK, III, p. 102); this has been noted also for algae in the tidal zone (see CLARK, 1917).

(ii) Diurnal Variations in Succulents

The above heading must be read very strictly "in relation to the pH of plant cells," because the general early history, as outlined by RICHARDS (1898), GUSTAFSON (1924–1925), SMALL (1929, pp. 82, 124–129), THOMAS (1951 a) and others, has now been expanded by an extensive literature concerned with what happens in what is now called "Crassulacean acid metabolism" (THOMAS, *op. cit.*, p. 162), in which pH values are still being given to the second decimal figure for expressed mixed juices of leaves that were shown in 1929 by LYNN (in SMALL, 1929, p. 127) to be differentiated in the pH of their tissues and also in the light-dark pH variations within these tissues. There are many interesting aspects of these much discussed phenomena, such as the enthusiasm about the large proportion of *iso*-citric acid which, however, has been shown by PUCHER *et al.* (1949) and again later by VICKERY (1952) to take no obvious part in the diurnal variations of organic acid content of *Bryophyllum* leaf: the variations being mainly in malic acid concentration with similar variations in the smaller quantities of citric acid.

The pH values of mixed sap quoted by THOMAS are—5.05 after 26 hours illumination at room temperature; 4.56 after the same period of light, followed by 24 hours darkness; 5.40 with light and dark periods as before but followed by 5 hours light at room temperature. GUSTAFSON (1925) got a variation from pH 5.37 at 4 p. m. on a bright day going down to pH 3.90 at 10 a. m. the following morning followed by a sharp rise to pH 4.43 two hours later (noon): on a dull day the variation was less but the fall was quicker, from pH 4.4 at 5 p. m. to pH 3.75 at 8 p. m. (see SMALL, 1929, p. 125). GUSTAFSON (*op. cit.*) followed the pH of sap from *Bryophyllum* kept in continuous darkness for 15 days from an initial value of pH 5.4 to pH 3.9 during the first day in darkness to pH 3.6 on the third day, up to pH 4.4 on the 6th day and pH 4.5 during the following 9 days.

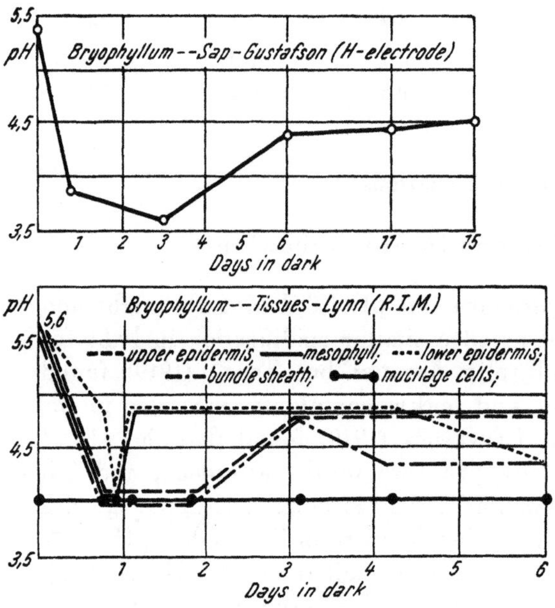

Fig. 2. Comparison of sap data (GUSTAFSON) and tissue data (LYNN), for *Bryophyllum* exposed to continuous darkness. (*ex* SMALL, 1929, p. 127.)

LYNN's 1929 data give the details for a six-days period of continuous darkness for tissues in *Bryophyllum* leaves. The daily variations were from pH 4.0 in the morning to pH 5.6 or less for all tissues except xylem, phloem, and a system of acid mucilage cells which were in hypodermal strips and·also some cells around the vascular bundles as parts of the bundle-sheath. These acid mucilage-cells remained at pH 4.0 day and night and during six days of continuous darkness: they have been suggested to be the location within the leaf of the similarly unvarying concentration of *iso*-citric acid (SMALL, 1954), which VICKERY (1952) states "may, in great part, be physically remote within the leaf-cells from the centres at which the major reactions are taking place;" the major reactions involving considerable variations in the concentrations of malic and citric acids. A comparison of GUSTAFSON's and LYNN's data was graphed for page 127 of SMALL's 1929 Protoplasma Monograph and this is here repeated as Fig. 2.

The biochemistry of the generalised phenomena as found in fresh juices of *Bryophyllum,* or in dried crushed and extracted material, has been discussed at length by BONNER (1950) and others, but there is still very little known about separate cells or tissues in this plant. Storage of *iso*-citric acid in quantity within special cells is not impossible, considering what is known for other storage products within idioblasts or even special tissues, and very small quantities, within the experimental error of estimations, would be available to complete any required intermediate link in the reaction chain.

(iii) Stomatal Guard-cells

The physiology of the guard-cells of stomata is still in a rather controversial stage. The phenomena of stomatal movements are numerous and diverse. There is a demonstrated variability of guard-cells even on one and the same leaf, so that factual evidence in support of any one explanatory theory can be obtained and contradicted by selection among observed facts concerning individual pairs of guard-cells. The writer has seen this conflict of evidence even in a single pair of guard-cells, one of which was red with BAN indicating a pH < 4.4, while the other was yellow indicating pH > 4.8. The starch content also varies, sometimes from one to the other of the same pair of guard-cells (see HEATH's beautiful photomicrographs, 1949).

WEBER's original theory of the control of stomatal apertures by a $CO_2 \rightarrow$ pH $\rightarrow$ enzymic activity $\rightarrow$ sugar concentration $\rightarrow$ turgor chain of controls (1923 b) is here set out as in the original so that the scope of later experimental work and theorising can be compared with that first logical development of a satisfactory hypothesis to cover the facts known in 1923 and many of the general average facts known in 1953.

„Es ist denkbar, daß die entscheidende Bedeutung, welche dem *Licht* für das Spiel der Spaltöffnungen zukommt, indirekt durch Veränderung des CO_2-Gehaltes des Blattinnern und damit der Wasserstoffionenkonzentration des Plastidenmediums der Schließzellen bedingt ist. Die Empfindlichkeit des Spaltöffnungsapparates für Schwankungen der Lichtintensität ist — wie aus den Untersuchungen LINSBAUERS hervorgeht — erstaunlich. Ein plötzlicher Wechsel in der Beleuchtungsstärke macht sich schon binnen 3—5 Minuten in einer Veränderung der Spaltenweite bemerkbar. Die Regulierung der Spaltweite ist so vollkommen, daß unter sonst gleichen Umständen am natürlichen Standort einer Pflanze ein bestimmter, der herrschenden Lichtstärke entsprechender Spaltöffnungszustand zu konstatieren ist."

„Wir können uns auf Grund der von ILJIN ermittelten neuen Tatsachen folgendes vorstellen: Bei veränderter Beleuchtungsintensität und damit zusammenhängender Schwankung der Assimilationstätigkeit — die ja, wie wir wissen, darauf ungemein prompt reagiert — erfährt der Säuregehalt der Schließzelle (vielleicht der Plastiden selbst) eine Veränderung; dies wirkt wiederum auf Intensität und Richtung der enzymatischen Vorgänge ein, bedingt weiterhin dadurch eine Verschiebung des osmotischen Wertes des Schließzellensaftes und auf diese Weise die Regulation der Spaltenweite. Es genügen in dieser Hinsicht jedenfalls minimale Verschiebungen der Wasserstoffzahl. ILJIN hebt ausdrücklich hervor, ,daß die Pflanze eine größere Empfindlichkeit als die üblichen Indikatoren ... zeigte; der Wasserstoffionengehalt mußte, um eine Farbänderung des Indikators herbeizuführen, im Vergleich mit der Reaktionsempfindlichkeit der Pflanze unverhältnismäßig erhöht werden'. Eine analoge Empfindlichkeit gegenüber dem aktivierenden Effekt zeigte auch das Proferment der Diastase bei den *in vitro* Versuchen BIEDERMANNS."

„Natürlich ist es nicht ausgeschlossen, daß das Licht auch in anderem Sinne auf den Ablauf der enzymatischen Prozesse innerhalb der Schließzellen einwirkt; entweder ebenfalls in indirekter Weise durch Änderung der Permeabilität der Plasmahaut oder aber in direkter Weise durch photodynamische Wirkung auf den enzymatischen Apparat."

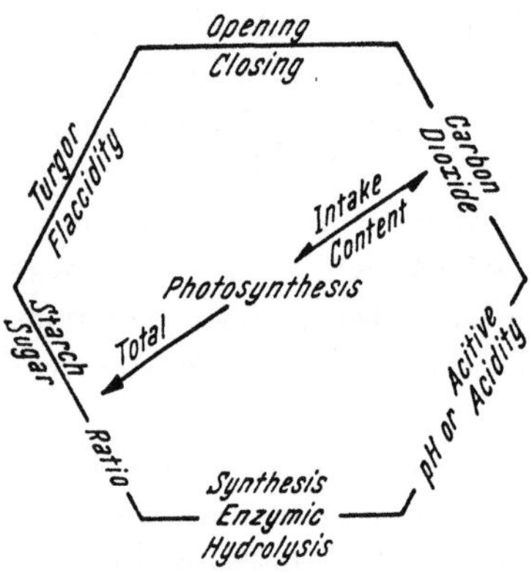

Fig. 3. Diagram of factors in stomatal movements. (After SMALL, 1947).

WEBER's enzymic control theory as expounded, with some more experimental evidence, by SMALL, MAXWELL, CLARKE, and CROSBIE-BAIRD (1939, 1942) may be summarised in "The Hexagon" which has been used three times already by SMALL, and also by two other authors in possible explanations of how stomata may on the average be opened and closed.

HEATH and WILLIAMS in a series of papers from 1948 to 1952 questioned the adequacy of this explanation on various grounds, one of these being concerned with the so-called rôle of starch but, as SMALL pointed out (1950), it is the concentration of sugar (below the maximum required to open the stomata fully) that is supposed to control the stomatal aperture. HEATH and MILTHORPE (1950), and also SCARTH and SHAW (1951 a, 1951 b), as well as several early workers, found that stomata *opened* in air free from carbon-dioxide as an external supply, but SCARTH and SHAW (1951 a) produce clear evidence that only about one fifth of the respiratory carbon dioxide is "captured by photosynthesis in CO_2-free air ..., even with light of 500 fc." and that in *Pelargonium* leaf photosynthesis (and pore *area*) increase with an increase from 50 fc. of light intensity even with external CO_2-free air. This indicates the presence of enough internal carbon dioxide for increased photosynthesis. It may be noted here that most stomatal physiologists refer their pore data to diameters or area in spite of proofs by SAYRE (1926), also by DIXON and BENNET-CLARK (1930), that the dimension which controls flow through an ellipse or a narrow slit is the perimeter, upon which the diameter of a narrow ellipse or slit has very little bearing.

Scarth and Shaw *(op. cit.)* conclude that their results are "consistent with the view that the main action of light on stomata operates through photosynthetic reduction of CO_2 concentration within the leaf." They suggest the presence af "a complex mechanism in the physiological response," but in their 1951 (b) paper state that later results confirm the above view concerning photosynthetic reduction of CO_2 content as a factor in stomatal opening.

Williams (1952) suggests a return to Weber's (1923 a) and Kisselew's (1925) theory of changes in protoplasmic permeability in response to light as controlling guard-cell turgor and stomatal aperture, but both Kisselew and Williams seem prepared to agree that carbohydrate changes may fulfil the useful function of reinforcing and stabilising changes brought about by other means. The introduction of permeability changes as a link in the chain of processes may account for *rapid* movements where stomatal closure takes less than 30 seconds, but this does not necessarily rule out the enzymatic control which can develop closure in one to two minutes and partial opening in three to five minutes (Small and Clarke, 1942, p. 250; Small, 1946, p. 95). Further, the effect of pH on the permeability of protoplasm is discussed with experimental evidence at some length by S. C. and M. M. Brooks (1941, pp. 236–243). As Small (1946) wrote—"It is clear that when problems involving permeability and pH are concerned, any possibility may theoretically be present." Cooke (1952) suggests that "A convenient explanation of many of the anomalous results in the literature of stomatal physiology would thus be afforded, *e. g.* the discrepancies between the results of various authors concerning the effect of artificially-produced pH changes on stomatal aperture."

It may be a relief to turn from the welter of theories and conflicting evidence to a correlation of pH values with one of the most generally accepted facts, namely that stomata do *tend* to be open during a sunny day and to be closed during darkness; coffee with its stomata closing around mid-day being a well-known exception (see Nutman, 1937 a, 1937 b, and Small, 1946, 1954). This correlation is demonstrated for sixteen species in Table V which is a systematic arrangement, according to R. I. M. pH-ranges, of data provided by Maxwell (1939) and Clarke (1942) for the expositions by Small. There are some differences between the ranges given for some of the species but the reproducibility of R. I. M. pH-ranges is evident—for two independent observers at an interval of three years under unspecified weather conditions. Earlier estimations of the natural pH values of guard-cells were made using tint methods, and the higher or lower values can be accepted only with many reservations: perhaps the different tints in physiologically similar guard-cells could yield *comparative* pH values, "higher" or "lower" as usually given with discreet vagueness as to actual numerical data.

Reports were given on the behaviour when immersed in four different buffer solutions at various pH values, of stomatal aperture and starch content of guard-cells in epidermal strips from 12 different species (Small and MAXWELL (1939) and from 6 more species (SMALL and CLARKE, 1942). Unsuccessful attempts to repeat these results with *Pelargonium* have been recorded by ALVIM (1949), and also by WILLIAMS and SHIPTON (1950), and there is, therefore, a difference of opinion concerning both the experimental facts

Table VI. *Natural pH of Stomatal Guard-cells in R. I. M. pH-ranges.*

	in Light	in Dark
Coffea arabica M	62–59	52–44
Coffea arabica C	56–56	40–40
Pelargonium zonale C	56–56	40–40
Rumex obtusifolius M	56–48	44–44
Rumex obtusifolius C	56–44	40–40
Vicia faba M	56–56	52–48
Vicia faba C	56–48	44–44
Chrysanthemum leucanthemum C	52–48	44–44
Centranthus ruber M	52–48	44–44
Centranthus ruber C	52–48	44–44
Iris pseudacorus M	52–48	44–44
Iris pseudacorus C	52–44	40–40
Kniphofia sp. M	52–48	44–44
Kniphofia sp. C	52–48	40–40
Inula helenium C	52-48	40–40
Hyacinthus sp. M	52–48	40–40
Hyacinthus sp. C	52–48	40–40
Tulipa sp. M	52–48	40–40
Tulipa sp. C	52–48	40–40
Tradescantia zebrina C	52–48	40–40
Narcissus pseudonarcissus M	52–44	40–40
Narcissus pseudonarcissus C	52–44	40–40
Bryophyllum calycinum M	44–44	40–40
Bryophyllum calycinum C	44–44	40–40
Beta vulgaris C	44–44	40–40
Rumex acetosa C	44–44	40–40

M = MAXWELL. C = CLARKE.

Revised and re-arranged from SMALL and MAXWELL (1939), SMALL and CLARKE (1942).

and the interpretation of the diverse observational data. The technique and the materials have varied which may explain some of the diversities; but physiology, particularly of stomatal guard-cells, tends to vary according to conditions, *e. g.* what guard-cells do under a cover-slip or porometer may be very different from what happens without a cover-slip or porometer, as is known very well to some investigators. The MAXWELL and CLARKE results were obtained under careful supervision, but CLARKE declared her results to be "all hay-wire," and individual observations were certainly mutually contradictory as stated in words and in tabular from in the original paper (*op. cit.*, p. 239). It was, as explained in the original, after SMALL had advised the averaging of each set of results for half-hourly periods ½ — 1 — 1½ — 2 — 2½ hours, that a strong negative correlation was demonstrated between *average* stomatal aperture and *average* starch content over the range of pH used for nearly all four buffers with six different species. In these circumstances it seems more than a little "naïve" to expect to be able to contradict a *general average* point of view by getting some more contradictory results with one kind of plant, one kind of buffer, at one particular pH value. The averaged results given by SMALL and CLARKE show definite peaks for stomatal aperture at certain pH values, sometimes more than one peak with the same material and the same buffer. There is no guarantee that the positions of these peaks are repeatable under different conditions with different materials and techniques, and the steepness of many of these peaks means that slight deviations in conditions could yield different results for aperture and/or starch content. This does not seem to be an appropriate place to discuss at any length the vagaries of guard-cells which are sensitive to many quite different types of conditions, including those summarised in The Hexagon and not excluding any one of the many others mentioned in the extensive literature, as for example the nuclear changes in stomatal guard-cells reported in some detail for *Vicia faba* by WEBER (1927) and again in more detail for a number of different species with different behaviour by SINKE (1939, p. 28–36, p. 85–87), and by COOKE (1952) for chloroplasts (see Chap. VII [b]).

7. Plant hormones

There are growth-promoting substances which are not acids (BER and MOSKWA 1951), but natural auxin and many other substances used for the control of growth are weak carboxylic acids although there is no foundation for a dogma on the subject such as — "One of the essential requirements is a side-chain containing a — COOH group."

A more or less general agreement has now been reached on one point, namely that for carboxylic acid growth regulators in general and for natural auxin in particular, the pH of the cells in which growth-hormone action occurs has a marked effect upon the activity or efficiency per unit of a given concentration of hormone. This connection was suggested by many investigators according to RIETSEMA (1950) who mentions BOROWIKOW, 1913; HOAGLAND, 1917; OLSON, 1923; LUNDEGARDH, 1923; and STRUGGER, 1926, 1928, 1932, 1934. RIETSEMA (*op. cit.*, p. 57) gives STRUGGER 1932, 1934, the credit of being the first to establish the connection experimentally, but immediately afterwards quotes VAN SANTEN (1940) as having repeated STRUGGER's experiments and shown that his results had been obtained with material in bad condition and adds that the sodium acetate buffers used by STRUGGER have a toxic effect on root growth, quoting (*op. cit.*, p. 60) COGGESHALL, 1931; SAKAMURA and KANAMORI, 1935; VAN SANTEN, 1938, 1940; and LUNDEGÅRDH, 1942; as having observed toxic action in a number of cases. With all this criticism, it seems better to consider BONNER's experiments on acid curvatures as at least one of the first clear studies of this subject (1934), although these too have been criticised as to the interpretation of results.

BONNER (1934) used pieces of decapitated *Avena* coleoptiles from each of which a 0.5 mm. strip of epidermis was removed along a narrow side. When these coleoptiles were immersed for 2 hours in McILWAINE's phosphate citric acid buffer solutions they showed acid curvatures of about 14.3° at pH 5.2, and about 63° at pH 4.1. Having demonstrated that extractable natural auxin was present in these pieces of coleoptiles, and having estimated colorimetrically the internal pH of crushed coleoptiles after immersion in solutions of various pH values, BONNER graphed his data and tabulated some of them. Bringing the relevant data into a Table with other calculated data SMALL (1946) gave a summary statement of these results which is useful.

Table VII.

pH of external buffers	2.5	3.4	4.1	4.7	5.7	6.4	7.0	7.2
Internal pH	4.7	4.8	5.2	5.7	5.9	6.1	6.3	6.9
pH for per cent change of curvature	4.8	4.9	5.1	5.5	5.9	6.2	6.6	7.0
per cent free acid (pK 4.65)	45	40	24	10	6	4	2	(1)
per cent change of acid curvature (means)	42	37	27	13	6	3	1	(1).
with internal pH	4.7	4.8	5.2	5.7	5.9	6.1	6.3	6.9

Modified from BONNER's 1934 Tables XV and XVI and Fig. 1.

The last three lines show how closely the auxin activity decreases with the fall in percentage of free acid as the pH is controlled at higher values which convert more and more of the auxin into the relatively inactive neutral salt form. These data concerning acid curvatures, as being large at pH 4.1, smaller at pH 5.2, and "little if any" at pH 7, can be readily accepted. They are in agreement with other results which demonstrate that auxin activity per unit is higher in the range pH 4.2–5.2 than in the range pH 6 to pH 7. The use of indicators on mixed juices has been deprecated *as a method of estimating the pH of cell-sap,* but in that very young relatively little differentiated organ the coleoptile, the pH of most of the cells may be the same within narrow limits, so that as a series of relative numbers BONNER's data seem to have some significance. The pK given by BONNER was 4.8; VAN SANTEN quotes pK 5.0 as that of natural auxin; but BONNER's 1934 titration curve cuts the 50 per cent line at pH 4.65.

VAN SANTEN (1938) using cylinders of *Avena* coleoptiles in phosphate buffer solutions obtained results very similar to those by BONNER. The data as tabulated in her Table I range from 14.1% growth in 24 hours with pH 3.75 to a peak at 23.9 per cent at pH 4.18 and then gradually down to 10.7 per cent at pH 7.00. The growth is rapid for the first four hours after which the growth-rate is low; and the growth-rate in the first 4 hours increases with pH decreases from pH 7.0 to pH 4.18, but falls at pH 3.75 to the same as it is for pH 5.33 (*op. cit.*, Fig. 4).

Rietsema (1950), like Bonner, Van Santen, and others found a 10 to 25 per cent growth increase in *Avena* coleoptile cylinders in 18 hours, the growth-rate increasing from 9.5 per cent at pH 5.9 to 25.6 per cent at pH 4.0, and also decreasing in another series from 23.8 per cent at pH 4.0 to 4.7 per cent at pH 7.7. In relation to the argument that auxin acts *only* when in the free acid condition, it may be noted that up to 6.8 per cent growth increase was found at an external pH 8.0. A later Table (*op. cit.*, p. 89) gives many similar results for growth in buffer solutions, with 0.2 mg./l indole-3-acetic acid added in 3, 8, and 24 hours: *e. g.* growth in 8 hours with 0.2 mg./l IAA is 25% at pH 4.5 and falls gradually to 10.4% at pH 6.7; but with 1.0 mg./l IAA growth in 8 hours varies somewhat irregularly and was 37.2% at both pH 4.1 and pH 6.3; other results on p. 92 are more regular.

Rietsema (*op. cit.*) used a micro-injection indicator method with R. I. M. interpretation of the *colours* seen, and claims that the pH of both vacuolar sap and cytoplasm remained in the 59–56 range, pH $< 5.9 > 5.6$, with bromo-cresol purple and methyl red, even after 24 hours immersion in 0.01 M phosphate buffer solutions of pH 4.2 and pH 5.8, but 62–59 or pH $< 6.2 > 5.9$ is given for BCP of indefinite colour, which may or may not be correct for an external pH 8.3. Rietsema (*op. cit.*) found that between pH 4 and pH 8 the rate of protoplasmic streaming was not affected by changes of external pH values. But Rietsema, who experimented and discussed over a wide range of phenomena, did *not* apparently make the logical suggestion that — if there is a pH effect on auxin activity and there is no apparent effect of external pH on the cytoplasm and vacuole, then the pH effect on auxin may occur *outside* the protoplast, in the cell-wall where and when the auxin is producing its recognised action on the extensibility of the cell-wall itself: this would not involve penetration of the cytoplasmic cell-lining or entry into the vacuole. Rietsema got as far as the outer layers of the cytoplasm being the location of pH effects on growth, and does not seem to concede the possibility of auxin action being primarily an effect upon the extensibility of the cellulose wall, which could then be a still more outer location than the one that he considers at some length. He was largely concerned with the problem of whether the free acid is active *only* in the non-dissociated condition, but this problem is not urgent in relation to our present subject — the pH of plant cells.

Some acidic effects upon the cell-wall are known, and Brecht (1936) after working on oats and sunflower pointed out that the similar action of acidic growth-substances need not be due to pH effects on protoplasm, nor need the action of pH changes be due entirely to their effects on growth-substances. The effects of both on the cellulosic wall must be considered. GESSNER found that increase of growth-substance and changes of pH down to pH 4.5 increased both elongation and plasticity (1934, 1936, see HEYŃ, 1940, for these and similar references). The positive evidence given by DIEHL *et al.* (1939) for direct effects of auxin on cell walls is even now better than any negative evidence which has been reported to the contrary. The position that auxin and other growth-substances can have effects only through the protoplast is not really tenable. Even if under natural conditions auxin is developed within the cytoplasm, it is known that it can pass from cell to cell and even out into agar by diffusion, so it is available for the demonstrated effects upon the extensibility of cell walls.

8. Chlorosis

There are many different kinds of chlorosis, and as yet the pH of plant cells seems to be known as an effective factor in only one of these deficiency symptoms, namely the deficiency of iron, but it may be a factor in other cases because its action is that of precipitation of iron when the high pH of cells surrounding the more acid iron-conducting tissues results in non-availability of iron in the normally chlorenchymatous cells beyond those in which the iron is precipitated. This chlorosis can be lime-induced as well as a condition which depends on an unusually high pH developed by other causes.

Effect on the pH of Plant Juices of Liming the Soil

A general rule is that the pH of plant sap is raised by liming the soil, and the pH of roots is affected more than that of stems. The buffering capacity of added lime comes into action as well as the alkalinity of the lime in the soil or soil-water. Exceptions to this general rule have been reported:— by BRYAN (1919) who found no change with maize or corn tops; by TRUOG and MEACHAM (1919) who found no change with lupines; and by CLEVENGER (1919) who stated that the shoot-sap pH may be lower with lime than without lime.

HAAS (1920) reported on many plants, with increases, decreases, and no change, varying from plant to plant, thus —

increase of pH on liming occurred in the juices of — alfalfa roots (6.06–6.12), tops (5.99–6.19); alsike clover roots (5.58–5.84), tops 5.28–6.19; barley tops 5.62–5.72; buckwheat seedlings 5.48–5.97; lupine tops 5.31–6.63; medium red clover roots 5.94–6.12, leaves 6.02–6.19, stems and petioles 5.63–5.95, tops 5.84–5.92; white mustard roots 5.62–5.91, entire plants 5.48–5.78; oat plants 5.65–5.67; timothy tops 6.12–6.19; winter wheat plants 5.95–6.33, leaves and stems 5.77–6.12.

decrease of pH on liming occurred in the juices of — alfalfa roots 6.21–6.12; field pea tops 6.80–6.53; garden bean tops 5.97–5.65; medium red clover roots 5.88–5.82; serradella tops 5.94–5.74.

no change in pH on liming occurred with corn tops 5.48–5.19, both limed and unlimed.

There was thus an increase on liming of 0.26–0.35 of a pH unit in juices from alsike clover roots, lupine tops, medium red clover stems and petioles, white mustard roots and entire plants, winter wheat leaves, stems, and whole plants; of 0.49 of a pH unit for buckwheat seedlings; of 0.91 for alsike clover tops; and a small increase of 0.06–0.20 of a pH unit for alfalfa tops and roots, barley tops, medium red clover roots, leaves, and tops, and timothy tops.

Decreases of 0.20–0.32 of a pH unit occurred with tops of field peas and garden beans, and with serradella tops: smaller decreases of 0.06–0.09 occurred with some alfalfa roots and with medium red clover roots.

No change of pH occurred with corn tops, and only a very slight increase of 0.02 of a pH unit with oats.

NEWTON (1923) using peas, beans, and barley, found that the pH of juices was not decreased by limiting the supply of calcium. DUSTMAN (1925) confirmed this for the tomato. The above results by HAAS (1920), however, indicate that liming the soil usually increases by 0.06–0.35 the pH of juices which have been expressed, mixed, and more or less freed from carbon dioxide by the hydrogen electrode technique: the exceptions are mainly some parts of leguminous plants, corn tops, and oats. But there is much more affected by the liming of the soil than the expressed juice of a plant.

LOEHWING (1930) investigated chlorosis of wheat plants, after having found (1928) that certain grain plants showed foliar chlorosis when grown on humus soils which were limed to correct soil acidity, and that this chlorosis increased after periods of intense sunshine or full insolation. Using expressed juice and a quinhydrone method which might be considered to cover any carbon dioxide effects, he found that wheat plants grown on limed humus soil and unshaded (eight weeks old) had their young leaves chlorotic and lacking in iron, while plants of this age on limed humus soil and shaded showed a lesser degree of chlorosis only after they were ten days older than eight weeks. ROGERS and SHIVE found something similar with maize (1932). LOEHWING also found that the juices of the tops of his wheat plants had pH values which varied according to the soil and the time of day or insolation, as well as with liming or no liming. His results are very instructive when tabulated arithmetically for nine-weeks-old plants (pH at eight weeks), rather than in a graph as he did for eight-weeks-old plants:

on humus soil — unlimed and unshaded, juice pH 5.50–5.85;
limed and unshaded, juice pH 6.12–6.67, limed plants chlorotic in 8 weeks with + 0.62–0.82 pH, and with nocturnal pH above 6.05;
unlimed and shaded, juice pH 5.47–5.61;
limed and shaded, juice pH 6.08–6.40, limed plants with some chlorosis after 9 weeks, with pH + 0.61–0.79 pH, with nocturnal pH just above 6.05;
on loam soil — unlimed and unshaded, juice pH 5.84–6.03;
limed and unshaded, juice pH 6.03–6.31, no chlorosis with + 0.19–0.28 pH, but with nocturnal pH just below 6.05, for limed plants.

The lowest pH values occurred mostly at 6 a. m., and the highest pH values occurred from 6 p. m. to 10 p. m.; between these upper and lower limits the pH value of the juice varied during the day.

From these results by Loehwing, it can be seen that liming humus soils raised the pH of plant juices by 0.61–0.82 of a pH unit; while liming loam soils raised the pH only by 0.19–0.28 of a pH unit. Liming humus soils resulted in chlorosis of wheat plants when the total pH of the juice was raised above pH 6.05 during the night, while liming loam soils did not have the same deleterious effect and did not raise the total pH value above pH 6.05. There would seem to be a critical value for chlorosis which is correlated with minimal or *nocturnal* pH values above 6.05 for the juice of the wheat plant. This question of the effect of pH and the lack of iron at the right place within the plant has been extensively studied. Loehwing found by microchemical examination that chlorotic plants contained an abundance of iron in the roots but little or none in young leaves. Olsen (1935, p. 27) gives a curve for iron in solution at different pH values in Knop's nutrient-solution, showing a sharp drop from 0.1 mg. iron per litre at pH 3.2, almost to zero at pH 6.0 (about 0.003 mg.), with about 0.002 mg. beyond pH 6.2 up to pH 9.0.

One of the most interesting and instructive applications of the differentiation in the pH of plant tissues yields an explanation of lime-induced chlorosis in the presence of sufficient iron in the soil. In any considerable mass of data a fresh viewpoint may disclose numerical correlations which have not been seen by the original producers of the data. This was the case above with Loehwing's data and the critical pH 6.05, which is obvious when his data are re-tabulated in the form given here. It is again the case for his 1928 data. I may be wrong about this pH 6.05, and Loehwing does not mention it at all, but the facts are as stated above, and also below.

A re-examination of Loehwing's 1928 data discloses the appearance of the same critical pH 6.05 (as for chlorosis) for extra iron-content in oats and maize. In four sets of results for oats, two show pH 6.2 and pH 6.5 with highest iron-content, and two out of four sets for maize also show highest iron-content for pH 6.1 and pH 6.2; whereas four wheat sets show no pH above 6.0 and no highest iron-content with highest pH. If Fe is the total iron-content and pH is the reaction of the expressed juice, Loehwing's data can be sampled to show that pH 6.1 or over is associated with extra iron, while pH 6.0 did not involve extra iron. The highest pH values in each set occurred when calcium carbonate was added to the nutrient medium.

					:			
Oats	Fe	0.046	0.053	*0.082*	:	0.081	0.097	0.061
	pH	4.9	4.5	*6.2*	:	4.7	4.4	5.8
Maize	Fe	0.036	0.045	*0.071*	:	0.091	0.093	0.070
	pH	5.6	4.7	*6.1*	:	4.8	4.2	5.8
Wheat	Fe	0.140	0.145	0.117	:	0.121	0.165	0.042
	pH	4.4	4.7	6.0	:	5.0	4.7	5.8

It had been observed by a number of workers, from HEYNE in 1815 onwards, that illumination tends to decrease the "sourness" or raise the pH value of plant juices. LOEHWING (1930) found that strong illumination, combined with liming of acidic base-deficient humus soil, induced chlorosis at an early stage in wheat plants: he also reported abundant iron in the roots, with little or no iron in chlorotic leaves. It has been said that iron begins to be precipitated from inorganic solutions at pH 3.5 and that precipitation increases to totality at pH 6.0 (see OLSEN, 1935, p. 27), with only organic compounds of iron remaining in solution above pH 6.0; but BRITTON (1942, II, p. 79) gives pH 2 as the point of precipitation for ferric iron and pH 5.5 as the point of precipitation for ferrous iron, as hydroxides. Reduction within the plant, of nitrates to nitrites, is well known, so that iron in solution in plants should normally be in the ferrous condition. These points must be borne in mind when considering the next stages of this problem and its solution.

SHIVE and INGALLS (1931) found, in a series of nine plants, *Bryophyllum*, *Rumex* (whole plant), buckwheat, *Sedum* (whole plant), tobacco, tomato, asparagus, soybean, clover, that the iron content of the leaves and some stems varied with the pH of the expressed juice in such a way that with increasing pH the *total iron increased* and the *soluble iron decreased*; but no correlation of lime-induced chlorosis and kind of iron-content was found by LINDNER and HARVEY (1944) using chlorotic leaves of various Rosaceous plants. SHIVE and INGALLS (1931) suggested a precipitation of iron, probably along paths of translocation, but "the mechanism by which this is accomplished is at present not clear." Precipitation at some distance away from the point of entry would lead to continued absorption and therefore to a higher total iron-content with a probable increase in the insoluble/soluble iron ratio. OSERKOWSKY (1932) found that the concentration of iron in the tracheal sap from chlorotic pear-trees was higher than that from green-leaved pear-trees.

The next stage was reached when ROGERS and SHIVE (1932), used the principle of SMALL's Range Indicator Method, mostly with different indicators, on fresh sections of various materials. Among other data, they reported accumulation of precipitated iron in tissues of relatively high pH (6.2–5.6) especially where such tissues lay adjacent to tissues of low pH, in a number of plants such as maize leaf and stem, clover petiole, and stems of soybean and potato. In chlorotic maize leaves, which contained "as much or more iron than do green maize leaves," the iron is confined to the bundle regions and penetrates, if at all, only to adjacent cells; chlorophyll in chlorotic leaves is confined to mesophyll cells adjacent to the bundles. Precipitated iron was also found to a small extent and uniformly distributed in the tissues of *Oxalis* stem and *Rumex* petiole, where the tissue pH values never reach above 4.4. In these cases the iron would appear to be in the ferric condition.

The above results by LOEHWING, and by ROGERS and SHIVE, seemed to have settled the explanation of lime-induced chlorosis. HAAS (1942) agreed that pH is a main factor, but LINDNER and HARVEY (1944) apparently do not think so. They do not quote ROGERS and SHIVE in their review of the literature [1], nor do they give any pH values for the material they used or in any other connection. They used extensive chemical analyses with fractions for solubilities in various media, and found that leaves affected with lime-induced chlorosis are *high in potassium* and low in that fraction of the iron soluble in 0.5 N HCl. No correlation of lime-induced chlorosis with any other fraction of the iron was found, such as those soluble in water, ether, alcohol, 1 N acetic acid, 1 N ammonium hydroxide, or 0.1 or 1.0 N HCl, or the electro-dialysable iron. Lime-induced chlorosis was also found to be associated with low values for calcium and magnesium. LINDNER and HARVEY suggest a replacement of iron by potassium "on the enzyme responsible for chlorophyll formation," leading to inactivation of the enzyme.

A re-investigation of their materials on the lines used by ROGERS and SHIVE might well clear up their problem. For example, their material consisted of lime-chlorotic leaves of pear, apple, peach, apricot, cherry, and wild rose, all of which belong to Rosaceae, one of the "all acid" families and so not strictly comparable with the cereals used by LOEHWING and with the maize, clover, soybean, and potato used by ROGERS and SHIVE. The Rosaceae material is probably more comparable with the acidic *Oxalis* and *Rumex* in which ROGERS and SHIVE found precipitated iron distributed uniformly about the tissues.

As similar dominantly humus soils are now beginning to be used as arable land in Britain, there is more than a purely American interest in experimental investigations of lime-induced chlorosis on humus soils. The appearance of this particular type of chlorosis during the cultivation of acid-peat land should not retard the cultivation of moors when the cause is properly understood. Since excess of potassium chloride leads to acidification of plant juices and liming sometimes leads to an undesirable degree of decrease in acidity, what seems to be needed is a suitable supply of both potassium and calcium, balanced so that the mixed expressed plant-juice does not reach the critical pH 6.05 and so that the leaf-mesophyll-cells retain an acidity below pH 5.6.

V. The Cell-Wall

In considering pH of plant cells, although this is a contribution to Protoplasmatologia, the cell-wall cannot be entirely ruled out: pH changes affect the cellulose wall and its activities and various maturing or differentiating processes affect the actual pH of the cell-wall. In this connection it must now be appreciated that according to modern views,

[1] Nor do REDISKE and BIDDULPH in their recent (1953) consideration of the translocation of iron, in which iron concentration is described as the "primary factor in determining the mobility of iron" in roots, with pH and available phosphorus as secondary factors but independent with simply additive effects.

pH is no longer merely a playground for theoretical physical chemists and that pH has become a more or less arbitrarily defined intrinsic property of all sorts of materials, particularly as a measure of base-avidities. WALLING (1950) on the acid strength of dry surfaces, states that "The most useful basis for the comparison of acid strengths of different media appears to be upon their ability to donate a proton to a neutral base, *i. e.* in terms of the H_0 function of HAMMETT and DEYRUP, defined as $H_0 = -\log a_{H^+} . fB . fBH^+$, where a_H is the activity of hydrogen ion and the f's the activity coefficients for a neutral (*i. e.* uncharged base), and its conjugate acid." This brings H^+ (hydrogen ion) rather than H^+H_2O (hydronium) into action, on a *dry* surface. WALLING (*op. cit.*) states also that the addition of H^+ *i. e.* a proton "is, or precedes, the rate-determining step in a great number of acid-catalysed reactions." He defines the acidity of a solid surface as "the ability of the surface to convert an adsorbed neutral base to its conjugate acid," especially in the case where the conversion involves the usual proton transfer. Indicators were used and a system of interpretation of the resulting colours was used along lines similar to those of SMALL's Range Indicator Method. Many *dry* substances were examined by WALLING, especially silica aluminas in connection with industrial catalytic cracking processes for petroleum, and an interesting datum was pH < 0.43 for dry copper sulphate. The reader may, with all due precautions, try the taste of the surface of a *crystal* of copper sulphate for some degrees of confirmation!

This modern view of pH was used by SMALL (1929) in his discussion of pH and the cell-wall (*op. cit.*, pp. 346 sqq.) where still older cases of its application to cell-walls were cited. ODÉN (1916) considered the pectins of cell-walls to be important not only as binding materials but also as pH regulators. PRIESTLEY (1924) stated that fats, fatty acids, and soaps can be detected in cell-walls from a very early stage and appear to be present as early as the cellulose and pectic substances regarded as typical of the plant cell-wall.

The effect of acids in bursting cells of various kinds (LLOYD and ÚLEHLA, 1926), even before they penetrate into the cell and coagulate the cytoplasm (ÚLEHLA and MORAVEK, 1922) was explained as due to adsorption of the hydrogen ion by the wall resulting in a sudden change of the degree or sign of the electric charge. As an example, the cytoplasm of *Basidiobolus* hyphae is almost or quite impermeable to strong acids, including .00005 N HCl which gives bursting of these hyphae in 30–60 seconds from first contact. Other materials used included pollen grains and pollen tubes of *Phaseolus odoratus* and apical regions of the algae *Trentepohlia, Cladophora, Vaucheria, Bornetia,* etc. The bursting of *Bornetia* before penetration of the acid into the vacuole was checked by means of a natural indicator in the cell-sap: but LLOYD and ÚLEHLA using methyl red as an introduced internal indicator found some penetration of strong acids, and came to the conclusion that the sudden bursting in acid "is directly connected with sudden changes in the biocolloids whether they compose the cell wall or the protoplasm."

The electrically neutral cellulose of the cell walls is not the only substance present, and the smaller quantities of other components have marked effects upon the pH phenomena of these walls. The general pH value of 52–48 for collenchyma and the common values of 44–44, 44–40, or 40–40 for lignified, suberised, or cutinised walls are normally due to solid phases rather than liquid phases in the cell-wall materials, but Ingold (Small, 1929) noted an acid oily liquid (pH 5.2) in the *intercellular* spaces of very young pericycle in sunflower hypocotyl. Mature lignified walls may begin as pH 56–56, pass to 52–48, and then become 44–44, 44–40, or 40–40, and even go to < 34. There are many records in the R. I. M. surveys of decreasing pH during lignification of cell-walls, particularly in the upper, middle, and lower parts of stems, *e. g. Veronica* and *Viburnum*, and some twigs of *Ligustrum* (Fig. 1). In sunflower (*Helianthus annuus*) the procambial strands are at 44–40 before lignification begins, but the xylem walls were noted as at 52–48 while unlignified, passing to 44–40 at the time of expansion of the cotyledons, and then to < 34 only at the mature stage in the stem below ground and for xylem fibres in the older part of the root at a similar age. In broad bean (*Vicia faba*) the procambial strands were less acid (52–48) but changed to 44–40 in the root immediately on germination and in the stem before the radicle was 4 cm. long: the lower pH < 34 was found among the fibres of xylem and pericycle near the base of the stem. The pericycle in the stems of other species is frequently lignified in parts with walls of fibres at 40–40 or < 34 while the cell contents of other pericyclic cells in the same region may be 44–40, 52–48, or even in the 59–56 pH range (cp. *Viburnum*). The pericycle in *Helianthus* stem is in the 59–56 pH range, except below ground where 44–40 is general after the expansion of the first foliage leaf, with < 34 in this region when the flowering stage has been reached; but 44–40 appears in the upper part of the root after the expansion of the second foliage leaf; and < 34 occurs in the pericycle of middle and all lower parts of the stem and upper parts of the root in the mature plant.

Suberisation of endodermal walls and in the formation of cork leads to pH-ranges like those of lignified walls, and Martin (1927 a) noted 52–48 for the cells cut off towards the outside from a cork cambium at 44–40, but these pH values were for cell-contents. Maturely suberised walls are < 34 as in *Sambucus* stem periderm, with 52–48 for phellogen cell contents: endodermal suberin is at 44–40 or only 44–44 in *Cerastium* stem, and < 34 is rare for endodermal walls. Cutinised walls are commonly in the 52–40 range, but < 34 is recorded for cuticle of the leaves of *Puya*, *Gasteria*, and *Rhododendron*: within the 52–40 range for cutin, 52–48 occurs for stems of *Reseda, Phyllocactus, Senecio, Lapsana*; 44–44 for stems of *Euphorbia* and *Aucuba*, also *Haworthia* leaf; 44–40 for stems of *Cheiranthus* and *Ligustrum*; and 40–40 for *Echinocactus* stem. Mature sunflower stem shows 44–40 in the outer cuticle as well as in the cell-contents of the epidermis, with an unstained cellulose inner layer to the outer cell-wall between these two coloured parts.

Armstrong (1929 c) found similar acidic virages, with upper and lower limits, in a series of 14 fatty acids, pelargonic, capric, lauric, myristic, palmitic, margaric, stearic, arachidic, and dihydroxystearic from the acetic series; tiglic, oleic, erucic, linoleic, and ricinoleic from the oleic series. These acids gave acidic virages when melted if solids, with CR, BTB, and BCP; also colours when dissolved in alcohol and precipitated with water in the cases of ER and MR. With the lower or more acidic indicators the virage varied, acidic with some and basic with others, just as was found for acid cell-walls: the pH values indicated included: pelargonic $<$ pH 2.2, tiglic 2.4 to 3.4, capric and oleic about 4.4, lauric and erucic in the 4.8–5.2 range, ricinoleic $<$ 4.4 with lower indicators not taken up.

Following the new outlook as exemplified by Walling (1950) and B. T. Brooks (1950), and also following Clark, who wrote (1928, p. 7) "Even beyond the measurable lie cases for which the presumption of ionisable hydrogen is often useful," we can assign to these fatty acids and to acid cell-walls certain pH values for comparisons of their degrees of base-avidity. Further, we can then attribute to these acidic cell-walls the properties and functions which are consequential upon their degrees of base-avidity as measured by apparent pH values, and use these properties in explanations of otherwise obscure phenomena [2], such as the differential elongation of cells in systems of conducting tissues, *e. g.* if auxin is more active in acid walls (at pH 44–40 as in sunflower procambial strands or at pH 52–48 as in the broad bean) than it is in the 59–56 pH range for the walls of the undifferentiated less elongated parenchymatous cells around these embryonic conducting cells, the conditions for differential elongation would seem to be maintained if not actually initiated by a differentiation in pH which, of course must be established by a preceding cause in the never-ending sequence of cause and effect. pH is not a transcendental cause of anything, but it is the interesting link with which this contribution is concerned, within a much wider and longer chain of biological sequences.

VI. Buffers in Plant Cells

It is customary to use buffers and buffer action in theoretical explanations of obscure physiological phenomena in order to deprecate the use of a pH change in such hypotheses. The word 'buffer' is used in a metaphorical psychological sense in many such arguments, without any clear idea of an arithmetical evaluation of the amount of buffer capacity required and with no idea at all of the actual buffer capacity available in the particular case being considered.

[2] Compare Caley and Stoffer (1953) who found pH changes over a range of as much as pH 5.1 to pH 10.2 and back to pH 6.5 during a mixing of barium chloride with potassium sulphate and attribute these changes to adsorption or desorption of hydrogen ions or hydroxyl ions on the surface of precipitates.

SMALL and his collaborators, MARTIN 1927–1928, INGOLD 1929–1930, ARMSTRONG 1929, JACKSON 1947, have recorded arithmetical quantities for a number of particular cases, giving analytical data as well as measured values of buffer-capacity in terms of VAN SLYKE's buffer index, β, or dB/dpH. SMALL has returned to this subject recently (1954), particularly on the soil-pH aspect but also from the changed point of view in relation to the re-definition of pH. Buffer-indexes have tended to be the Cinderellas of pH expositions, with hydrogen-ion-*activity* to keep them company until 1950. SMALL (1929) gave rather lengthy arithmetical expositions of buffers and buffer-indexes in plants, using his knowledge of the few particular measured cases in plants then available. There has been very little extension of this work so far as plants are concerned, but buffer indexes have at last received suitable mathematical exposition in RICCI's (1952) systematic treatment of pH, buffers, and buffer capacity, although he has changed β into π, π a, π b; β having been used for something else. Since π is incurably associated with 3.1415926536 etc. and β spells "buffer index" to the present writer, VAN SLYKE's β = buffer index will continue to be used throughout this contribution.

Buffer systems consist of a bimolecular mixture of two substances with a one-hydrogen-ion exchange possible within each system. An addition or subtraction of hydrogen ions is compensated or buffered by a change in the proportions of the two substances with only a very slight change in the pH of the mixture. On the addition of NaOH to the system HAc + NaAc the Na^+ and Ac^- ions tend to balance, while the H^+ and OH^- ions combine to form water, and the pH rises only slightly because of the increase of the ratio of (Ac^-) over undissociated (HAc), so long as there is a reserve of undissociated HAc. The capacity for buffering of a buffer system depends on the concentration of the system, (undissociated acid + salt) or (acid-salt + neutral salt) or for a tribasic acid (more-acid salt + less-acid salt), and the concentration of each system *as a whole* is at its maximum, for any given molar concentration of the two substances together, when their proportion is 50 : 50: when the proportion is 20 : 80 or 80 : 20 the concentration of the *system* is only 20 = 40 per cent of its maximum for a given molality.

Characteristics of buffer systems depend upon (1) the concentration of the buffer system as a whole which depends on the weight, the pH, and the degree of dissociation of both components; and (2) the point or *range* on the pH scale of the maximum buffer action of the system, which is around the pH corresponding to the pK or dissociation constant for the acid or acid-salt concerned. The buffer action of a mixture of buffer-systems is completely and clearly additive over any one short range of pH which may be 0.2 or 0.5 of a pH unit in practice, as SMALL emphasised repeatedly in 1929 (see RICCI, 1952, for a systematic mathematical treatment of simple and complex and various intermediate systems, with illustrative buffer index curves). Theoretically there is only one main *shape* for the buffer-index curve of any one buffer system covering effectively about three pH units, but its position on the pH scale is governed by the degree of dissociation, the pH corresponding to the pK being the position of the maximum of 0.376 for 1.0 molal concentration. This maximum, of course, varies with total concentration, being .0576 for 0.1 M, .00576 for 0.01 M, .000576 for 0.001 M, and so on; but the shape of the curve remains the same if drawn on an appropriate scale.

In plant cell-saps the individual values of the component systems of the total buffer complex cannot be precisely formulated, but MARTIN (1927) found that in 0.2 steps over the range pH 5.6 to pH 4.2 the buffer capacity of sunflower hypocotyl sap was completely accounted for by the .0038 to .0067 M phosphate found by analysis. These buffer values were measured in terms of MILROY's phosphate ratios. MARTIN (1928) found by analysis of sap from broad bean stems and roots about .0134 M phosphate, .011 M malate, .021 M oxalate, and .005 M bicarbonate, which when added together in terms of β values were not more than .0015 β higher than the sap β values obtained by titration and subsequent calculation. The formula is simple:—

$$\frac{(\text{molarity} \times \text{vol. used in mils) of acid or base}}{\text{vol. in mils of sap} \times \text{pH shift}} = \frac{dB}{d\text{pH}} = \beta$$

INGOLD (1930, p. 442) found the β value for potato tuber sap to be .0105 β from pH 5.0 to pH 6.0, which is higher than the β value .0068 calculated from the known components of the buffer complex: these were .0057 M citrate, .005 M malate, .0053 M phosphate, .0003 M oxalate, 0.3 per cent asparagin; and a heat coagulum, probably tuberin (0.5 per cent). MILLER (1931) reported a β value of .023–.036 for pH 5–6, and GUTHRIE (1934) gave a higher citrate content .013–.027 M for American tubers. The higher American buffer indexes are accounted for by the higher citrate content to some extent, but GUTHRIE also found that his analytical data were not enough to account quantitatively for his higher buffer-index values, so there is at least one still unknown buffer system in potato tubers.

ARMSTRONG (1929 a) obtained buffer-index values for saps of ten kinds of fleshy fungi: maximum β values varied from .0018 β for *Polystictus versicolor* to .0275 β for *Armillaria mellea.* The buffer complex of *Collybia velutipes* was analysed in detail and the additive values for the components accounted for the whole of the buffer index curve for the sap of this fungus, thus:—

maximum β value pH 4–5 = .0257

pH ranges	4.0–4.5	4.5–5.0	5.0–5.5	5.5–6.0	6.0–6.5	6.5–7.0
$\frac{\beta \text{ of buffers}}{\beta \text{ of sap}}$	97%	101%	103%	107%	89%	85%

The analytical results were:— .0228 M malate, .014 M citrate, .010 M phosphate.

ARMSTRONG (1929 b) also explained the buffer index curve for sap from young aerial stems of *Pelargonium* with a fit within the limits of experimental error. The initial pH was 4.2 with β from 4.0–4.2 = .0644 which compared with β 4.0–4.2 for the analysed buffer complex at .0613 or 95%. The analytical results were .0573 M oxalate, .0337 M malate, .0066 M citrate, .003 M tartrate, .010 M phosphate, and .0019 M metallic anhydride.

INGOLD (1930 a) gave buffer index curves for potato tuber and leaf, turnip, onion bulb, carrot, beetroot; and in (1930 b) he gave buffer index curves for a number of organic acids and for juices of rhubarb, lemon, potato, apple, and bean stem. The malate, oxalate, tartrate, amino-acetate and succinate data for the buffer-index curves given on p. 363 of SMALL's 1929 monograph were obtained, as stated (*op. cit.*, p. 357) by INGOLD using a quinhydrone electrode on 0.025 M solutions: the other curves in the same figure were calculated from standard Tables and data in the literature. If *suitable* titration data are published β curves can readily be calculated from any detailed data of titration curves.

JACKSON of SMALL and JACKSON (1949 a, 1949 b) obtained buffer-index curves for the fresh expressed saps from roots of barley, oats, and wheat, from turnips and potato tubers, and related these and other analytical results to the ecological tolerances of these plants, on the basis of a theory of ecological soil-pH tolerances used by SMALL (1946) in a classification of plants from that point of view. These data are here presented as Table VIII.

Table VIII.

Plants	Oat	Potato	Barley	Wheat	Turnip
part	root-sap	tuber-sap	root-sap	root-sap	tap-root sap
malate	.0288 M	.0046	.0054	.0038	trace
citrate	.0020 M	.0037	.0036	.0009	—
phosphate	.0115 M	.0032	.0039	.0020	.0052
initial pH	6.1	5.9	6.0	6.1	6.3
relative β values	β .0425 = 100				
at pH 4.6–4.8	100	47 β .02	34	15	9
at pH initial	26	22	11	7	5
at pH 6.4–6.6	26	17	*10* β .004	6	6
Tolerances	amphi-tolerant	amphi-tolerant	alka-tolerant	meso-philous	meso-philous

The buffer capacity, measured as buffer index or β values of root cell saps, may have a real value to plants and be of some agricultural and ecological importance, as well as being of academic physiological interest. Amphi-tolerance appears to be connected with β .02 or more around pH 4.7 and β .004 or more around pH 6.5: alka-tolerance with β .004 or more around pH 6.5 with β below .02 around pH 4.7: mesophilous below both these apparently operative values, below .02 β around pH 4.7 and below .004 β around pH 6.5. SMALL (1954) describes and discusses buffers much more fully than can be done in this contribution which is confined to plant cells.

VII. The Protoplast

The term "protoplast" is used to cover the semi-permeable limiting layer; the hyaline cytoplasm with all its inclusions; nucleus with chromosomes; chloroplasts and other plastids; and all the granules of various other kinds which make up a very heterogeneous living protoplast. Whether the hyaline cytoplasm is always neutral or not, it is quite certain that some of these inclusions are not always neutral, so there is something to be said about their pH values and the effect of pH upon them.

(a) The Nucleus and Chromosomes

ZIRKLE and BLOOM (1953) have shown that the chromosomes themselves are the parts of the protoplasts that are most sensitive to bombardment by protons or hydrogen ions, as distinct from the hydronium (H^+H_2O) ions which are normally concerned in pH. Bombardment by 2-Mev protons was localised to an 80 per cent degree within a circle 2.5 μ in diameter and many well-known qualitative effects were observed by means of time-lapse photomicrography. These effects included temporary and permanent chromosome bridges, partial and complete inhibition of anaphase and metaphase movements and of constriction or cleavage, and displacement of constriction; leading to unequal distribution of chromosomes into daughter nuclei, and other phenomena. No detectable effects were observed in the extrachromosomal regions, in the cytoplasm, or the centriolar regions. Therefore it is to be expected that pH changes will also have their most obvious effects upon the chromosomes, their appearance and behaviour.

The nucleus with its chromosomes has not only been bombarded and broken up irreversibly, it has been subjected to many chemical and physical reversible treatments by SINKE (1939) and others. SINKE's summary of the literature is very extensive and detailed, but his own experiments have very logically led him to the conclusion that the pH effects observed by KUWADA and SAKAMURA (1926), and SAKAMURA (1927) are only particular examples of the hydration and dehydration of chromosomes which can be attained in many ways other than by the experimental control of the pH of the nuclear medium by its carbon dioxide content. SINKE (*op. cit.*) used many different plant materials and obtained results which are consistent within themselves, although his material showed great variation even within one set of apparently equivalent cells, such as the stomatal guard-cells of one species. Using direct observation of individual nuclei under the microscope, with both ordinary and dark field illumination, SINKE (*op. cit.*) has confirmed KUWADA and SAKAMURA's (1926) conclusions that in *Tradescantia* the chromosomes are reversibly made more distinct in the range pH 3 to pH 5, and less distinct above pH 10, but he used epidermal cells from young petals of *T. reflexa* whereas the previous investigators were working with pollen-grain mother cells of *T. virginica*, and found hydrational indistinctness at pH 6.6. SINKE found that the nuclei he observed "showed no marked change" and that "cytoplasmic streaming was active" with $^1/_{500}$ M NH_4OH (pH 9.8). With $^1/_{100}$ M NH_4OH (pH 10.4) "the chromonemata were slightly less distinct," and the cytoplasmic streaming was "somewhat inactive."

SINKE (*op. cit.*, p. 114–115) found considerable variation in living nuclei, and classified them on the basis of their reactions with hypotonic and hypertonic solutions into homogeneous (with refractivity of chromonemata near that of karyolymph) and heterogeneous (with chromonemata more highly refractive than karyolymph and therefore more easily seen). He subdivided the homogeneous nuclei into three types: (1) chromonemata strongly hydrated, homogeneous even during artificial dehydration, no detailed structure at any time; (2) showing structure (heterogeneity) during artificial dehydration; (3) karyolymph strongly dehydrated, refractivity near to that of chromonemata: dehydration of types (1) and (2) yields type (3). On the basis of reactions to heating above 60° C., SINKE divided the *heterogeneous* type into two sub-types; with heat treatment (a) retains its heterogeneity with distinct chromonemata, while (b) becomes homogeneous with very obscure chromonemata or none visible.

The relevance of five different types of behaviour, for chromosome threads and nuclear sap (karyolymph) in relation to reversible dehydration-hydration phenomena, is obviously as great for pH as for other factors; and no one kind of behaviour can now be claimed as *typical*. When chromonemata and karyolymph have different refractivities they can be distinguished and the chromosomes are clearly visible. Variations of any kind, including the pH of karyolymph, chromonemata, and surrounding cell-sap, may produce differential dehydration or hydration and it is a matter for experimental observation, rather than theoretical generalisation, as to what actually happens with any given material under particular conditions.

The pH values noted in connection with visibility (differential dehydration) of chromosomes were pH 3.8 for distinct appearance and pH 6.6 for indistinctness in pollen-mother-cells of *Tradescantia virginica* (KUWADA and SAKAMURA, 1926), with complete reversibility: also *(ibid.)* distinctness from pH 2.6 to pH 5.2, with swelling above pH 5.2 definite at pH 5.4, and chromosomes "scarcely to be distinguished" at pH 6.7. At the same time pH 7.0 is indicated for the pressed juice of the anthers of the same plants. In *Secale cereale* anthers (SAKAMURA, 1927) the internal pH varied from meiosis (5.8–6.0) to the generative nuclear division in young pollen grains (5.2), and the chromosomes were visible without CO_2 transfusion.

The control of intracellular pH was by buffer solutions (KUWADA and SAKAMURA, 1926), by carbon dioxide content (SAKAMURA, 1927), and by very dilute acetic acid or ammonia (SINKE, 1939), who also found similar nuclear changes taking place spontaneously in the guard-cells of stomata, as WEBER (1927) observed earlier. In open stomata the guard-cell nuclei of *Vicia faba* were reported by WEBER as dense and refractive, while those of closed stomata were completely hyaline, in the sol condition. These same appearances were seen by SINKE in *Vicia faba, Tradescantia virginica, Zebrina pendula,* and some other plants (*op. cit.*, p. 28): and he also reported distinct chromonemata in the guard-cell nuclei of half-open stomata; but he warns that these changes are to be seen only in a limited number of plants. No regular sequence of changes was seen by SINKE "in *Sagittaria sagittifolia, Podophyllum pleianthum* (?), *Ceratopteris thalictroides,* and some other plants;" although he gives a detailed, timed, account of reversible changes induced by saccharose solutions, 0.4 M and 1.6 M, during 70 minutes in *Zebrina pendula* guard-cells. He quotes an earlier paper (SINKE, 1937) for many similar observations.

With all the observed variations of the chromatin components of the nucleus, particularly those in which pH is a factor, it is naturally to be expected that there will be some variations in the records for the general pH of the nucleus as a whole. These records are mainly concerned with the colours given by the nucleus with various indicators on different plant materials. There is SCHAEDE's (1924) observation with onion bulb scale epidermal cells where the nucleus and cytoplasm were yellow with methyl red (> 5.6) while the sap was red (< 5.2): injury induced a red with MR (< 5.2) in the cytoplasm. COLLA's (1928) data do not include a separate item for the nucleus, but the protoplasm was less than pH 5.9 (see SMALL, 1929, p. 304). SMALL's summary statement concerning the results from the R. I. M. survey included notes concerning pH 52–48 for injured or dead nuclei (red MR = < 5.2, yellow BAN = > 4.8); above 5.6 for living nuclei (yellow MR = > 5.6): either red (< 5.6) or yellow (> 5.9) with ethyl red; never more than a faint mauve usually yellow with BCP (> 6.2 or < 5.9); and never green or blue with BTB (< 6.4). This leads to about pH 5.9 or pH 5.6 for living plant nuclei under ordinary conditions, with pH < 5.2 > 4.8 for nuclei of injured cells. The upper limit seems to be < 6.4 with BTB never blue or green, only yellow. The lower limit for normal living plant cell nuclei may not be pH 5.6, but it may be about pH 4.4 with BCG always green (> 4.4) in living cells. COOKE (1952, Table V) was able to distinguish at 4.15 a. m., using BAN, a reddish (<4.8) nuclear area and definitely yellow chloroplasts in the guard-cells of stomata in *Endymion hispanicus,* while with MR both chloroplasts and the rest of the guard-cells were uniformly red (< 5.2): this red with MR was maintained throughout the day for the rest of the cell, (while the plastids became various shades of yellow, from dull to vivid), so that whether the nucleus was included among the plastids and went up from its early morning < 4.4 to > 5.6 or whether it remained < 5.2 with the "rest of the guard-cell" it remained yellow with BCP (< 5.9) around noon

(12.30 p. m.) and was usually red with ER (<5.6), so that 56–44 is the extreme range possible, with the lower limit checked at BAN red <4.4, BCG green >4.4. The red colour with BAN is, however, not necessarily that of an uninjured nucleus: COOKE (1952, p. 135–136) describes four stages in the development of BAN colours in *Endymion* guard-cells. This indicator is toxic and causes shrinkage of these guard-cells before the differential red staining of the nucleus occurs. This differentiation may be due to SINKE's dehydration phenomena and not directly to any acidification of the chromonema by the BAN toxicity. As usual there is great variation and sometimes colour differentiation cannot be seen.

MILOVIDOV (1949) notes SCHAEDE's (1924) above-mentioned results and, without any mention of results obtained by KUWADA and SAKAMURA (1926), SMALL (1929) or SINKE (1939) as given above, gives other data which seem to be in general agreement with the above account of chromosomes rather than whole nuclei. According to MILOVIDOV's summary (1949, p. 263) animal nuclei have been recorded within the range pH 6.4 to pH 7.8 or pH 7.1 ± 0.7, but plant nuclei have been found usually within the range pH 4.0–5.8 or pH 4.9 ± 0.9. MILOVIDOV quotes STRUGGER, 1932, as finding that in the onion nuclei may be pH 3.83–4.04 when young but 4.4–4.49 when old: this claims a second decimal accuracy which is not justifiable under the new outlook on pH, and these results should be taken as indicating about pH 4.0 when young and about pH 4.4–4.5 when old. LEHOTZK, 1936, got pH 5.6 for nuclei from the same source, which may have been alive and uninjured (MILOVIDOV, 1949). YAMAHA (1932, *ex* MILOVIDOV, *loc cit.*) is quoted as giving pH >5.0 for the nucleolus, pH 4.6–4.8 for "Gerüstkaryotin" and pH 4.0 for "ganze Kerne" of isolated tapetal cells from the anthers of *Lilium tigrinum.* All the other results quoted in MILOVIDOV's Table of pH values for plant nuclei *(loc. cit.)* are from YAMAHA and ISHII (1933), and all these data were obtained for nuclei which had been isolated from the containing cells. These authors seem to have been more interested in kataphoretic isoelectric points of the nuclei which they claim as having survived isolation ("überlebenden" or "frei lebenden") while they rejected those which were "anscheinend getötet." They mounted these isolated nuclei, and also plastids, in various solutions usually 0.1 N KCl, and added 0.1 N HCl to change the pH value of the medium. Kataphoresis under about 5–10 milliamps was observed microscopically, and the IEP was deduced from a change in the direction of movement. The pH of the IEP was designated pHi, following REISS's suggestion (1926, p. 113/14) in his discussion of the importance of the relation between IEP and pH. He emphasised the physico-chemical neutrality of ampholytic substances around their own isoelectric points and not at pH 7.07.

While the physico-chemical IEP of isolated "überlebenden" nuclei can be accepted as being somewhere near the truth for these particular protein substances, their medium inside the living cell usually has a pH value well above the IEP, as SAKAMURA, REISS and others have emphasised. The pH effects vary with the difference between pHi and pH, and it has been to the variation in these differences that most attention has been directed, although SMALL (1918–1923) theorised on possible effects of changing the sign of the charge by alterations of pH values. Since the IEP of the relatively more acid plant proteins is usually low (pH 2.5–4.2), and the sap is normally in the range pH 4.0–6.2, it follows that changes of the sign of the charge are rare.

All these pH effects, however, cannot get rid of the fact that YAMAHA and ISHII's data for 'natural' pH values are for *isolated* nuclei. Most of these values would show red with methyl red for nuclei, and this has not happened, over a very wide and prolonged experience with many different kinds of materials. There are some cases, but when so many cells show yellow with methyl red, a red nucleus would be conspicuous in a yellow cell: pH < 5.2 is not typical of plant cells.

The actual pH values of nuclei given by YAMAHA and ISHII (1933) were obtained by using tints (Farbenton), mainly of BCG, and tints of green or blue with bromo-cresol green can all be classed as pH > 4.4, but the relative development of the blue element in the dichroic colour depends on many other factors as well as on pH, so that the tint, even if repeated with very similar material, is no certain guide to pH values of nuclei or other cell-parts such as plastids, even when these are living within the complete cell.

YAMAHA and ISHII (1933, p. 209) have a different point of view, they say— „Nach alledem ist höchstwahrscheinlich, daß der pH-Wert des frei überlebenden Zellkernes annähernd der wahren Reaktion des lebenden Zellkernes entspricht." In their Table 19 they give pH data for cytoplasm (Z), nuclear sap (K), and chromosomes (C) of pollen-mother-cells in *Tradescantia reflexa* at various stages of nuclear division thus:— early prophase pH 4.6–4.8 ZKC; later prophase pH 4.4 ZC, 4.6–4.8 K; metaphase, anaphase, and telophase pH 4.6 Z, pH < 4.0 C; pollen-tetrads pH 6.2 ZK. Similarly they record in *Tinantia fugex* var. *erecta* early prophase pH 5.0–5.2 ZK; later prophase pH 4.4–4.6 Z, pH 5.0–5.2 K; metaphase pH 4.4–4.6 Z, pH < 4.0 C; anaphase and telophase pH 5.0–5.2 ZC; pollen-tetrads pH 6.2 ZK.

SCHAEDE (1924) and SMALL (1929) described nuclei of injured cells as at pH 5.2–4.8, with nuclei of uninjured cells pH 5.6–5.9. The process of isolation may be the cause of some or most of the low pH values for nuclei reported by YAMAHA and ISHII (1933) but even these values show variations from tissue to tissue of the same plant, as well as differences at various stages of nuclear division.

For isolated nuclei they reported pH values as follows:—

for *Allium cepa* — epidermis of bulb-scales 4.2, green leaf cells 4.2–4.4, white leaf cells 4.6, bulb-scale mesophyll cells 4.6–4.8, root cells 5.0–5.2; with cell-sap 4.8–5.0 except for 4.0–4.2 in epidermis;

for *Impatiens balsamina* stem — outer cortex 4.8, xylem 5.0, pith 5.6, phloem 5.8; with cell-sap varying from 4.0 in epidermis, 4.6 xylem, 5.0 in cortex and phloem, and to 5.6 in pith: *op. cit.*, Table 18.

The stem pith cells of *Stapelia gigantea* had nuclei at pH 4.4 with cell-sap at pH 4.2, while the stem pith cells of *Pelargonium zonale* had nuclei pH 4.2 with cell-sap at pH 5.4. The pH range 4.4 to 4.9 covers the variation found for isolated nuclei of *Tinantia fugex* var. *erecta* 4.4, *Vicia faba* 4.4, *Tradescantia reflexa* 4.5–4.9, *Allium fistulosum* 4.6, *Lilium tigrinum* 4.7, also *Gastrodia elata* stem cells 4.9.

Whether or not YAMAHA and ISHII's isolating technique results in some acidification by injury, their technique has given many *differential* results, both for nuclei from different tissues of the same kind of plants and for nuclei at different stages of division. There is, therefore, either variation in natural pH values for nuclei and their component parts or variation in their degrees of acidification under uniform treatment or both; and all these values are below the lower limit of pH 6.4 in MILOVIDOV's collection of pH data for animal cell nuclei. The need for separate accounts of plant and animal cells in relation to pH is very obvious.

(b) Chloroplasts

There is a large modern literature about chloroplasts, but the idea of estimating the intrinsic pH of these essentially green, more or less solid, bodies seemed to be rather hopeless from both the electrometric and colorimetric points of view. There is no previous reference to this subject known to the writer, except YAMAHA and ISHII's (1933) *isolated* chloroplasts, but he cannot allow to pass this opportunity for putting on record some observations made by one of his research students, Miss M. P. COOKE (1952).

COOKE was investigating various aspects of stomatal physiology, when she recorded for epidermal strips of *Endymion hispanicus* with BAN "guard-cells red at first, then colour changed to orange yellow in the protoplasm, red in the nuclear area, and yellow in the chloroplasts": this applied to guard-cells in the light with normal air, and also with carbon-dioxide-free air both in light and darkness; as well as to guard-cells in normal air and darkness about 2 p. m. except that the peripheral protoplasm was then pinkish in its final stage. At the same time with MR she noted "guard-cell chloroplasts brilliant yellow, rest of cell uniformly red" for CO_2-free air in light, and "some guard-cell chloroplasts yellow, others orange, rest of cell red" for CO_2-free air in darkness; for normal air in light with MR "guard-cell chloroplasts dull yellow, rest of cell red," and for the same in darkness "guard-cells all red, including chloroplasts."

With BAN yellow $= > 4.8$ and red $= < 4.4$
MR yellow $= > 5.6$ and red $= < 5.2$

these records indicate for these particular chloroplasts

in light, air normal	> 4.8	> 5.6
in light, air CO_2-free	> 4.8	> 5.6
in dark, air normal	> 4.8	< 5.2
in dark, air CO_2-free	> 4.8	? 5.6 or less.

So that in light the chloroplasts were indicated as above pH 5.6, and in darkness as in the 52–48 pH range for normal air at least.

Much more work was done on this aspect of stomatal physiology, and it was confirmed that with BAN these chloroplasts were yellow under the various conditions used (> 4.8), and that with readily penetrating MR they were yellow to orange except in darkness with normal air when indefinite or red or dull yellow occurred (red = < 5.2, yellow = > 5.6). In particular, CO_2-free air and darkness left the chloroplast colour yellow with MR.

BPB gave no very definite indications with these chloroplasts, but BCG was green throughout the whole day (> 4.4) and any colour given by BCP was yellow (< 5.9), while ER colours developed in 2 minutes but were apparently reduced to some leucoform very quickly, being absent (colourless) in less than 15–20 minutes. The MR colours with chloroplasts varied under natural conditions during the day, being red (< 5.2) like the rest of the guard-cell at 4.15 a. m.; orange or red in separate cells at 5.15 a. m.; orange or mostly yellow (> 5.6) at 6.30 a. m.; mostly bright yellow (> 5.6) with some orange at 7.45 a. m.; vivid yellow (> 5.6) with rest of cell red (< 5.2) 9.00 to 10.50 a. m.; yellow with some orange (also yellow with BCP) at 12.30 p. m.; mostly orange with some yellow (> 5.6), a few pink, and rest of cell red (< 5.2) from 2.00 p. m. to 3.35 p. m.; mostly orange with none definitely red at 4.45 p. m. (to 9.50 p. m.). The BAN colour remained yellow (> 4.8) from 4.15 a. m. to 4.45 p. m.; at 5.15 p. m. the colour was indefinite: but at 8.50 p. m. the chloroplast colours with BAN were red (< 4.4) to orange (indet. ? between 4.8 and 4.4). (Cooke, 1952, Table V. Date 19. 5. 1952. 55° N, 6° W).

So that the guard-cell chloroplasts' pH began as 52–48 at 4.15 a. m.; was mostly above or around pH 5.6 from 6.30 a. m. to 12.30 p. m.; and began to fall below pH 5.6 in some guard-cells from 2.00 p.m. onwards; until at 8.50 p. m. the pH was in most guard-cells below pH 4.8 (? < 4.4). On two occasions the chloroplasts gave yellow with ER (> 5.9), otherwise they were colourless, or red (< 5.6) if examined quickly.

The MR timed records in Cooke's Table V do not go beyond 4.45 p. m. but elsewhere (*op. cit.*, p. 145) she notes that "the chloroplasts were again red by 8.50 p. m." as at 4.15 a. m. These guard-cells were examined under low-power magnification and also under a $^1/_{12}$" oil immersion objective with careful manipulation of the intensity of illumination in order to see the chloroplast colours as distinct from those of the rest of the guard-cell. The 12-hours record with methyl red differentiation in time, combined with a longer period for BAN colours, for chloroplasts alone, demonstrates that the original Iljin-Weber hypothesis (p. 81) covered another possibility which has been found experimentally: it included „vielleicht der Plastiden selbst" among the possible variations of pH in the guard-cells.

Cooke's (1952) observations include some on *Pelargonium* leaf in which guard-cells that appeared orange with BAN under low-power were found under an oil-immersion lens to have red "protoplasm" and yellow chloroplasts. Differences in distribution of chloroplasts within the guard-cells affect the general colour as seen under low-power; those of *Pelargonium* being comparatively large and uniformly distributed. Many of Cooke's results still await publication but these items give new information which should be known in any consideration of pH in connection with the inclusions of the protoplast. It should also be mentioned that with methyl red variations the observed normal associations are yellow (>5.6) with open stomata and red (<5.2) with closed stomata and intermediate tints with intermediate sizes of stomatal apertures (COOKE, 1952, p. 142).

(c) Granules

Having considered differential pH phenomena in the larger bodies of the heterogeneous protoplast, namely the nucleus with its homogeneous or heterogeneous variously undifferentiated or differentiated chromonemata, and the chloroplasts with their natural diurnal variation demonstrated in at least one plant, we should be quite prepared to accept, without too much theoretical objection, more or less any records of pH differentiation amongst the smaller granules of the protoplast. HEILBRUNN (1928) was inclined to emphasise the hyaline cytoplasm as being the main essential component of protoplasm and in 1952 he was still apparently of a somewhat similar opinion, for he then wrote (HEILBRUNN, 1952, p. 67) about SPEK and his work "But in his studies he has been concerned chiefly with granular (or really vacuolar) inclusions, and his data, for the most part at least, do not concern the hyaline protoplasm itself." The whole of this contribution and most of the work done on the cell is in a similar position, largely because heterogeneity is a characteristic with considerable interest and a living stability, whereas homogeneity is essentially unstable in that every differentiation is a subtraction from its homogeneousness and threatens a carefully balanced condition.

Thus there is a living interest, whether or not the data are due to minute granules, in the phenomena next mentioned in HEILBRUNN's masterly summary of work on the pH of protoplasm (1952, pp. 59—68), namely a variation from pH 1.6 to pH 8 during the life cycle of the myxomycete *Physarum polycephalum* as shown by its natural indicator, according to SEIFRIZ and ZETZMANN (1935). Earlier (*loc. cit.*) HEILBRUNN mentions SPEK as considering "that the pH of various protoplasmic constituents varies from 5 to 8," and this is something like the point of view of the present writer except for pH 5 as the extreme lower limit: There is some evidence for the possibility of chloroplasts sometimes being reddish with BAN (pH <4.4), and the *surface* pH of dehydrated chromonema may well be below pH 5. The evidence presented by SEIFRIZ and ZETZMANN (1935) is accompanied by a definite colour scale for the natural indicator and is at least as good for the range pH 1.6 to pH 8 (orange through yellows to green) as is the evidence with tints of artificial indicators and micro-injection given for *Amoeba* by SPEK and CHAMBERS (1933): it seems better, because of the use of tints in the earlier work where the *tints* of the CLARK and LUBS indicators were said not to harmonise in a series. There is no reason why tints should do so: an R. I. M. colour interpretation of the *Amoeba* colour data leads to about the same value pH >7.0 to pH 7.4 as is given for *Amoeba dubia* by

Spek and Chambers, but that does not disprove the conclusion of Seifriz and Zetzmann: and in both cases the material is cytoplasmic. Gutstein (1932) grew his material, yeast and various bacteria, in culture media containing very diluted indicators of many kinds, and obtained colours which indicate about pH 6.2 for yeast, *Staphylococcus* and *Streptococcus*, as well as pH 7.2–7.6 for other kinds of bacteria. There is no reason to disbelieve his results either. The organisms actually multiplied so that, in the concentrations used, the indicators were far from being fatally toxic.

All this leads up to the point that apart from its larger granules and inclusions, so-called hyaline cytoplasm is full of smaller granules in a great range of sizes, granules which are readily seen with dark ground illumination, some of which are so small that they exhibit lively Brownian movements. If these were coloured by indicators they would give definite colours to the material. So that if we continue downwards to molecular size, some protein molecules might be coloured while others were not coloured, and the molecules to which indicator dyes were attached would join the group of heterogeneous parts of the protoplast: so also would the limiting layer so ably described in great detail by Kopaczewski (1933), and S. C. Brooks (1941 and 1939).

Since the hyaline characteristic of cytoplasm seems to depend on what instruments and techniques are used for viewing the cytoplasm, the present writer sees no reason for excluding from valid pH values of cytoplasm any of the observed data, including Colla's (1928) colours which give for yellow with BCP < 5.9 and for living cytoplasm either yellow > 5.9 or red < 5.6 with ER, in *Chara crinita*; and Small's (1929) data for the acid epidermal cells of potato eyes where the BCP yellow indicated pH < 5.9, and the ER yellow indicated pH > 5.9, with MR checking at yellow $= > 5.6$, so that the cytoplasm there, with micromanipulated puncturing of cells coloured by immersion in indicator solutions, was demonstrated as at pH 59–59 or about pH 5.9.

Many cytoplasmic reactions are surface phenomena and the new outlook on pH as a surface characteristic applies to all the many fluid or solid surfaces known as heterogeneities in the cytoplasm whether that looks hyaline or not with some particular instrument and some particular illumination.

As Sinke and others have seen, the living cytoplasm may be hyaline at one pH, but opaque and quite visible at another pH value of the surrounding medium. As can be seen with dark ground illumination and various oil-immersion objectives granularity ranges form large plastids down to micelles; and with other techniques may reach lower sizes of "granules."

In the R. I. M. surveys and in later work on stomatal guard-cells (Cooke, 1952) the indicator colours were frequently observed to fill the plant cells *completely*, thus extending at least to the surfaces of the living cytoplasm, including its microscopic and ultramicroscopic granules, so that acidity of at least the surfaces of plant cytoplasm in the range pH 6.2 to pH < 4.8 is regarded as established beyond any reasonable doubt. Although Small's

1929 observations on the acid epidermal cells of potato yielded his only record of cytoplasmic pH (5.9) for the cytoplasm of *acid cells* (< 5.2), the hundreds of records in the R. I. M. survey where the cell-contents indicate 56–56, 59–59, or 56–59 are records of cytoplasmic pH values within the range pH 5.6–5.9 because of the colours involved; *i. e.* red sap obscures yellow in the peripheral cytoplasm but yellow sap would not obscure red or purple in peripheral cytoplasm.

(d) The Limiting Layer

The actual pH of the cytoplasmic membrane or limiting layer is that of a surface, a hydrophilic system with its degree of hydration depending on the ions with which it is in contact (KOPACZEWSKI, 1933). The interpretation of the resulting phenomena, which include the permeability of this surface or layer, should be brought up-to-date by a re-consideration in the light of the modern outlook on the pH of surfaces.

Under the new outlook, pH is a measure of what may be called "base avidity" as a term of wider application than concentration or activity of hydrogen (H^+) or hydronium (H^+H_2O). The theories of base avidities under non-aqueous conditions have still to be worked out, but the practical uses of a knowledge of grades in base avidities are quite extensive, including the acid-catalysis of oil into petroleum and other products. This phenomenon of ionic activity at surfaces has received some attention, *e. g.* in the Appendix to Chapter III of ADAM (1941) "Physics and Chemistry of Surfaces"; "it appears possible that the only satisfactory definition of pH in the surface may be one which varies rapidly at different depths" (*op. cit.*, pp. 411/12). WALLING (1950) in his definition of acid strength supports this point of view (see "The cell-wall," p. 93 above).

With these ideas in view it is easy to see the importance of heterogeneity of surfaces and their *surface* pH values throughout the interior of the protoplast, particularly in relation to the rates of enzyme action; and there are many phenomena concerning the permeability of the limiting layer which are more readily understood when it is known that the pH of the surfaces, either the outer surface of the protoplast or the surface of the medium with which that is in contact, may be able to "donate a proton" and so produce changes intimately related to the pH values of the surfaces concerned. ILJIN (1928), for example, concluded that the pH of the external solution has a great influence on the permeability of the protoplast. In general, the degree of permeability is lowest around the neutral point, increasing with increase or decrease of pH: but the minimum varies in its position on the pH scale with the ions and molecules permeating. This is to be expected when circumstances determine the direction in which protons are being donated. The experimental or natural conditions require careful definition in all observations concerning permeability since theoretically any possibility may occur and be reversed.

SMALL (1929, pp. 301–324) dealt at some length with these matters, enzyme activity, viscosity, staining with acidic or basic dyes, permeability, equilibrium points or zones as affected by buffer systems rather than isoelectric points for protoplasts, and membrane buffering; all from the old points of view concerning pH. Now there is the full account by Brooks and Brooks (1941) of data and views up to 1939: S. C. Brooks (1939) is in that bibliography as in "process of publication." But the 1950–1954 outlook on surface pH values and their significance has still to penetrate into biological observational investigations and theoretical considerations before any real summary can be given. Base avidities have been used for a long time, *e. g.* in connection with cell-walls, following Clark, who wrote (1928, p. 7) "Even beyond the measurable lie cases for which the presumption of ionisable hydrogen is often useful"; but surface acidities are now beginning to receive the attention they require, after a period of empirical commercial application of the mostly unknown principles involved.

Bibliography

Adam, N. K.. 1941: Physics and Chemistry of Surfaces. 3rd Edit. O. U. P.

Allen, R. C., 1943: Influence of aluminium on the flower color of *Hydrangea macrophylla* DC. Cont. Boyce Thompson Inst. **13**, 221–242.

Alvim, P. de T., 1949: Studies on the mechanism of stomatal behaviour. Amer. J. Bot. **36**, 781–791.

Armstrong, J. I., 1929 a: Hydrogen-ion Phenomena in Plants I. Hydrion Concentration and Buffers in the Fungi. Protoplasma **8**, 222–260.

— 1929 b: Hydrogen-ion Phenomena in Plants II. An investigation of the Buffer Complex of Sap from stems of *Pelargonium* sp. Protoplasma **8**, 313–343.

— 1929 c: Hydrogen-ion Phenomena in Plants III. The acidity of certain cell-walls considered in relation to the higher fatty acids. Protoplasma **8**, 508–521.

Arrhenius, S., 1887: Über die Dissociation der in Wasser gelösten Stoffe. Z. physik. Chem. **1**. 631.

Atkins, W. R. G., 1922: The H. i. c. of plant cells. Sci. Proc. Roy. Dublin Soc. (N. S.) **16**, 414.

— 1923: H. i. c. of the soil in relation to the flower colour of *Hydrangea hortensis*. Sci. Proc. R. D. S. **17**, 23.

Baird, J. G., and J. H. Prentice, 1930: Changes with age of pH of egg white and egg yolk. Analyst **55**, 20.

Bartholomew, E. T., 1923: Internal decline of lemons II. Growth rate, water content, and acidity of lemons at different stages of maturity. Amer. J. Bot. **10**, 117–126.

Bates, R. G., 1948: Definitions of pH scales. Chem. Rev. **42**, 1–61.

— 1954: Electrometric pH Determinations. Theory and Practice. 331 pp. New York and London.

Bell, R. P., 1952: Acids and Bases. London.

Ber, A., and Moskwa, 1951: The influence of organic solvents on the growth of plants. Experientia **7**, 136–137.

Biedermann, W., 1921: Das Koferment, Komplement der Diastase. Fermentforschung **4**.

-- Die organische Komponente der Diastasen. Fermentforschung **4**.

Bjerrum, N., 1923: Die Konstitution der Ampholyte, besonders der Aminosäuren und ihre Dissoziationskonstanten. Z. physik. Chem. **104**, 147.

Blank, F., 1947: The Anthocyan Pigments of Plants. Bot. Rev. **13**, 241–317.

Bonner, J., 1934: The relation of hydrogen ions to the growth rate of the *Avena* coleoptile. Protoplasma **21**, 406–423.

— 1950: Plant Biochemistry. New York.

Boyle, R., 1664: Experiments and considerations touching colours. London.

Brecht, F., 1936: Der Einfluß von Wuchsstoff- und Säurepasten auf das Wachstum von *Avena*- und *Helianthus*-Keimlingen und seine Abhängigkeit vom Sauerstoffgehalt der Luft. Jb. wiss. Bot. **82**, 581.

British Standard 1647: 1950: British Standards Institution, 24/28 Victoria Street, Westminster, London S.W. 1.

BRITTON, H. T. S., 1942: Hydrogen Ions. Vols. I and II. London.
BROOKS, B. T., 1950: Catalysis and Carbonium Ions in petroleum formation. Science **111**, 648–650.
BROOKS, M. M., 1926: Penetration into *Valonia* of oxidation-reduction indicators; estimation of the rH of the sap. Proc. Soc. Exp. Biol. and Med. **23**, 265.
— 1930: The pH and rH of the sap of *Valonia* and the rH of its protoplasm. Protoplasma **10**, 505.
BROOKS, S. C., 1938: The Chemical Nature of the Plasma Membrane as revealed by Permeability. American Naturalist **72**, 124–140.
— 1939: Ion Exchanges in Accumulation and Loss of Certain Ions by the Living Protoplasm of *Nitella*. J. cellul. a. comp. Physiol. (Am) **14**, 383–401.
— and M. M. BROOKS, 1941: The Permeability of Living Cells. Protoplasma-Monographien Nr. 19. Berlin.
BRYAN, O. C., 1919: Effects of different reactions on growth and nodule formation of soy beans. Soil Science **8**, 227.
BUXTON, B. H., 1929 a: The pH value of cell sap of flowers. J. Roy. Hort. Soc. **54**, 1–11.
— and F. V. DARBISHIRE, 1929 b: On the behaviour of "anthocyanins" at varying hydrogen-ion concentrations. J. of Genetics **21**, 71.
BUXTON, B. H., 1932: Genetics of the primrose *Primula acaulis*. Jour. Genetics **25**, 197–205.
CALEY, E. C., and R. L. STOFFER, 1953: Changes in concentration of hydrogen ion during precipitation reactions between neutral salt solutions. Science **118**, 749–750. (18. 12. 1953.)
CHAMBERS, R., and T. KERR, 1932: Intracellular hydrion concentration studies. Jour. Cellular and Comp. Physiology **2**, 105.
CHENERY, E. M., 1937: The problem of the blue hydrangea. J. Roy. Hort. Soc. London **4**, 568–569.
— 1948–1950: Contributions to the bio-geochemistry of aluminium. Col. Off. (1529) 51. Agr. Res. Stn. Uganda.
CLARK, L., 1917: Acidity of marine algae. Pujet Sound Marine Sta. Publ. **1**, 22.
CLARK, W. M., 1928: The Determination of Hydrogen. Ions. 3rd Edit. London and Philadelphia: also 1st edit. 1920 and 2nd edit. 1922–1927.
CLEVENGER, C. B., 1919: H. i. c. of plant juices II. Soil Scence **8**, 227.
COLLA, S., 1928: Nota sulla azione della conc. d. H. i. sulle correnti protoplasmatiche. Protoplasma **5**, 179.
COOKE, M. P., 1952: Some aspects of stomatal physiology. Thesis. Q. U. B.
CROZIER, W. J., 1916: Cell penetration by acids. Jour. Biol. Chem. **26**, 217, 235.
— 1919: Intracellular acidity in *Valonia*. Journ. Gen. Phys. **1**, 581.
CZAPEK, F., 1922–1925: Biochemie der Pflanzen. Dritte Aufl. Jena.
DENNY, F. E., and W. J. YOUDEN, 1927: Acidification of unbuffered salt solutions by plant tissue Amer. J. Bot. **14**, 395.
DIEHL, J. M., C. J. GORKER, G. VAN ITERSON and A. KLEINHOONTE, 1939: The influence of growth hormone on hypocotyls of *Helianthus* and the structure of their cell walls. Rec. d. Trav. Bot. Néerlandais **36**, 709–798.
DIXON, H. H., and T. A. BENNET-CLARK, 1930: The Stomatic Control of Transpiration. Nature **126**, 601.
DOYLE, J., and P. CLINCH, 1928: Further notes on the metabolism of conifer leaves. Proc. R. I. A. **38**, B, 116.
DUSTMAN, R. B., 1925: Inherent factors related to absorption of mineral elements by plants. Bot. Baz. **79**, 233.
GAUDICHAUD, M., 1848: Des sucs séveux acides, et de quelques excrétions alcalines. Compt. Rend. **27**, 37.
GUSTAFSON, F. G., 1924: Total acidity compared with actual acidity of plant juices. Amer. J. Bot. **11**, 1–6.
— 1925: Diurnal changes in the acidity of *Bryophyllum calycinum*. J. Gen. Physiol. **7**, 719.
— 1927: Chemical Analysis of Tomato Fruits. Papers Mich. Acad. Sci. Arts and Letters **8**, 121.
GUTHRIE, J. D., 1934: Metabolism of citric, sulphuric and nitric acid in the potato tuber. Cont. Boyce Thompson Inst. **6**, 247–268.
GUTSTEIN, M., 1932: Bestimmung der H-Konzentration in der lebenden Hefe- und Bakterienzelle. Protoplasma **17**, 454–470.

HAAS, A. R. C., 1916 (a): The acidity of plant cells as shown by natural indicators. J. Biol. Chem. **27**, 232.
— 1916 (b): A simple and rapid method of studying respiration by the detection of exceedingly minute quantities of carbon dioxide. Science **44**, 105.
— 1920: Studies on the reaction of plant juices. Soil Science **9**. 341.
— 1942: Lime-induced chlorosis in relation to Soil Factors. Plant Physiology **17**, 27–51.
HEATH, O. V. S., 1949: Studies in stomatal behaviour. II. The role of starch in the light response of stomati. New Phytologist **48**, 186–211.
— and F. L. MILTHORPE, 1950: Studies in Stomatal Behaviour. V. The role of carbon dioxide in the light response of stomata. Jour. Experimental Bot. **1**, 29–62, 227–243.
HEILBRUNN, L. V., 1928: The Colloid Chemistry of Protoplasm. Protoplasma Monographs No. 1. Berlin.
— 1952: An Outline of General Physiology. 3rd Edit. Philadelphia and London.
HEMPEL, J., 1917: Buffer processes in the metabolism of succulent plants. Compt. Rend. Lab. Carlberg **13**, 1.
HERRMANN, H., 1879: Handbuch der Physiologie.
HEYN, A. N. J., 1940: The Physiology of Cell Elongation. Botan. Rev. **6**, 515–574.
HEYNE, B., 1815: On the desoxidation of the leaves of *Cotyledon calycina*. Trans. Linn. Soc. **2**. 213.
HURD, A. M., 1923: Acidity of corn and its relation to vegetative size. Jour. Agric. Res. **25**, 11.
— 1924: The course of acidity changes during the growth-period of wheat. Jour. Agric. Res. **27**, 725.
HURD-KARRER, A. M., 1928: Changes in the buffer system of the wheat phant. Plant Physiology **3**, 131–153.
ILJIN, W. S., 1914: Die Regulierung der Spaltöffnungen im Zusammenhange mit der Veränderung des osmotischen Druckes. Beih. Bot. Zentralb. **32**, 15–35.
— 1915: Die Regulierung der Spaltöffnungen. Beih. Bot. Zentralb. **32**, 1. Abt.
— 1922 (a): Wirkung der Kationen von Salzen auf den Zerfall und die Bildung von Stärke in der Pflanze. Biochem. Zeitschr. **132**.
— 1922 (b): Synthese und Hydrolyse von Stärke. Biochem. Zeitschr. **132**.
— 1922 (c): Physiologischer Pflanzenschutz. Biochem. Zeitschr. **132**.
— 1928: Die Durchlässigkeit der Protoplasmas. Protoplasma **3**, 558.
INGOLD, C. T., and J. SMALL, 1928: H. i. c. of plant tissues. IX. Improved technique for the R. I. M. Protoplasma **3**, 458.
INGOLD, C. T., 1929: H. i. c. of plant tissues. X. Buffers of the potato tuber. Protoplasma **6**, 51.
— 1930 (a): Hydrogen-ion Phenomena in Plants. IV. Buffers of Potato (tuber and leaf). Protoplasma **9**, **441–446**.
— 1930 (b): Hydrogen-ion Phenomena in Plants. V. The buffer systems of plant juices. Protoplasma **9**, 447–455.
— 1930 (c): Hydrogen-ion Phenomena in Plants. VI. Apparatus for measuring the effect of carbon dioxide on the reaction of plant sap. Protoplasma **9**, 456–458.
IRWIN, M., 1919: Comparative studies on respiration. VI. Increased production of carbon dioxide accompanied by decrease of acidity. J. Gen. Physiol. **1**, 399.
JACKSON, T., 1947: The buffer complex in the sap of some plants of economic importance. Thesis. Q. U. B.
JACOBS, M. H., 1920: The production of intracellular acidity by neutral and alkaline solutions containing carbon dioxide. Amer. J. Physiol. **53**, 457.
— 1922: The influence of ammonium salts on cell reaction. Jour. Gen. Physiol. **5**, 181.
KAPPEN, H., 1918: Untersuchungen über Wurzelsäfte. Landw. Vers. Stat. **91**, 1–40.
KAPPEN, and M. ZAPFE, 1919: Die Azidität der Pflanzensäfte unter dem Einfluß einer Kalkdüngung. Landw. Vers. Stat. **93**, 135.
KIESEL, A., 1930: Chemie des Protoplasmas. Protoplasma-Monographien **4**. Berlin.
KISSELEW, N., 1925: Veränderung der Durchlässigkeit des Protoplasmas der Schließzellen im Zusammenhange mit stomatären Bewegungen. Beih. Bot. Zentralbl. **41**, 287–308.
KOLTHOFF, I. M. trans. N. H. FURMAN, 1926: Indicators. New York and London
KOPACZEWSKI, W., 1926: Les Ions d'Hydrogène. Paris.
— 1933: La couche limitante cellulaire. Protoplasma **20**, 407–439.
KRAUS, G., 1883: Die Acidität des Zellsaftes. Abhandl. Naturf. Ges. Halle **16**, 154.

KROTKOF, G., D. G. WILSON, and R. W. STREET, 1951: Acid metabolism of McIntosh Apples during their development on the tree and in cold storage. Canad. Journ. Bot. **29**, 79–90.
KUWADA, Y., and T. SAKAMURA, 1926: A contribution to the colloid chemical and morphological study of chromosomes. Protoplasma **1**, 239.
LAPICQUE, L., 1921: Influence des acides et des bases sur une algae d'eau douce. Compt. rend. soc. biol. **84**, 493.
— 1922: Mécanisme des échanges autre la cellule et le milieu ambiant. C. R. Ac. Sci. **174**, 1491.
— and M. KERGOMARD, 1923: Acidification par l'asphyxie chez les Spirogyres. C. R. Soc. Biol. **88**.
Letter Circular LC 993, 1950: United States National Bureau of Standards, Department of Commerce, Washington, 25. D. C., U. S. A.
LEWIS, G. N., 1912: The activity of the ions and the degrees of dissociation of strong electrolytes. J. Am. Chem. Soc. **34**, 1631.
LILLIE, R. S., 1909: On the connection between changes of permeability and stimulation and on the significance of changes in permeability to carbon dioxide. Amer. Jour. Physiol. **24**, 14.
LIMPEROS , G., and K. E. RANTA, 1953: A rapid screening test for determination of the cholinesterase activity of human blood. Science **117**, 453–455.
LINDNER, R. C., and C. P. HARLEY, 1944: Nutrient interrelations in lime-induced chlorosis. Plant Physiology **19**, 420–439.
LLOYD, F. E., and V. ULEHLA, 1926: The rôle of the wall in the living cell. Trans. Roy. Soc. Canada **20**, 45.
LOEHWING, W. F., 1928: Calcium, potassium and iron balance in certain crop plants in relation to their metabolism. Plant Physiology **3**, 261.
— 1930: Effects of insolation and soil characteristics on tissue fluid reaction in wheat. Plant Physiology **5**, 293–303.
Marconi Instruments. Ltd., 1951: Hydrogen Ions - pH Measurement with particular reference to the Glass Electrode. St. Albans.
MARTIN, S. H., M. W. REA and J. SMALL, 1926: The Reaction of plant tissues. Report Oxford (94th) meeting Brit. Assoc. Adv. Sci.
MARTIN, S. H., 1927 (a): H. i. c. of plant tissues. III. Tissues of *Helianthus annuus*. Protoplasma **1**, 497.
— 1927 (b): H. i. c. of plant tissues. IV. Buffers of sunflower hypocotyl. Protoplasma **1**, 522.
— 1928: H. i. c. of plant tissues. VII. Buffers of sunflower stem and root. Protoplasma **3**, 273.
— VIII. Buffers of bean stem and root. Protoplasma **3**, 282.
MAYER, A., 1875: Über die Bedeutung der organischen Säuren in den Pflanzen. Landw. Versuchsstationen **18**, 428.
— 1878: Über die Sauerstoffausscheidung einiger Crassulaceen. Landw. Versuchsstationen **21**, 277.
MCCLENDON, J. F., 1914: On the electric charge of the protoplasm and other substances in living cells. Zschr. f. physik.-chem. Biol. **1**, 159.
MICHAELIS, L. (Trans. W. A. PERLWEIG), 1926: Hydrogen Ion Concentration. Baltimore.
MILLER, L. P., 1931: The effect of treatments with ethylene chlorhydrin on the pH of the expressed juice of potato tubers. Contr. Boyce Thompson Inst. **3**, 321–336.
MILOVIDOV, P. F., 1949: Physik und Chemie des Zellkernes. Erster Teil. Protoplasma-Monographien, Band 20. Berlin.
MISLOWITZER, E., 1928: Die Bestimmung der W. i. k. von Flüssigkeiten. Berlin.
NEWTON, J. D., 1923: A comparison of the absorption of inorganic elements and the buffer systems of legumes and non-legumes. Soil Science **15**, 181.
NUTMAN, F. J., 1937 (a): Studies of the physiology of *Coffea arabica*. I. Photosynthesis in coffee leaves under natural conditions. Ann. Bot. N. S. **1**, 3, 353–367.
— 1937 (b): Studies of the physiology of *Coffea arabica*. II. Stomatal movements in relation to photosynthesis. Ann. Bot. N. S. **1**, 681.
— 1938: Stomatal movement and epidermal water-content. Nature **141**, 608.
ODÉN, S., 1916: Zur Frage der Azidität der Zellmembranen. Ber. d. deut. bot. Gesellsch. **34**, 648.
OLSEN, C., 1935: Iron absorption and chlorosis in green plants. C. R. Lab. Carlsberg **21**, No. 3, 15–52.

Oserkowsky, J., 1932: H. i. c. and iron content of tracheal sap from green and chlorotic pear trees. Plant Physiology **7**, 253.
National Bureau of Standards, U. S. A., 1950: Standardisation of pH measurements made with the glass electrode. Letter Circular. LC 993 (supersedes LC 933).
Palitzsch, S., 1916: Sur l'emploi de solution de borax et d'acide borique dans la détermination colorimétrique de la concentration en ions hydrogène de l'eau de mer. C. R. Lab. Carlsberg **11**, 199–211.
Pantin, C. F. A., 1923: Determination of the pH of microscopic bodies. Nature **111**, 81.
Pfeiffer, H. H., 1925 (a): Eine Methode zur kolorimetrischen Bestimmung der W. i. k. in pflanzlichen Gewebeschnitten ohne Anwendung von Moderatoren. Zeitschr. f. wiss. Mikroskopie **42**, 396–414.
— 1925 (b): Über die W. i. k. als Determinationsfaktor physiologischen Gewebegeschehens in der sekundären Rinde der Pflanze. New Phytologist **24**, 65.
— 1926: Der gegenwärtige Stand der kolorimetrischen Azidimetrie in der Gewebephysiologie. Protoplasma **1**, 434–465.
— 1927: Über Unterschiede im Chemismus der Trennungsgewebe bei periodischem und Frostlaubfall usw. Botanisches Archiv (Mez) **18**, 319.
— 1940: Experimentelle Cytologie. Chronica Botanica Co.
Priestley, J. H., 1924: The fundamental fat metabolism of the plant. New Phytologist **23**, 1.
Pucher, G. W., H. B. Vickery and C. C. Sherman, 1946: The determination of citric acid in biological material. Jour. Biol. Chem. **113**, 235–245.
— *et al.*, 1948: Studies in the metabolism of Crassulacean plants: Diurnal variation in organic acids and starch in excised leaves of *Bryophyllum calycinum.* Plant Physiology **24**, 610.
Rea, M. W., and J. Small, 1926: The H. i. c. of plant tissues. II. Flowering and other stems. Protoplasma **1**, 3, 334.
— 1927 (a): The H. i. c. of plant tissues. V. The tissues of *Vicia faba.* Protoplasma **2**, 1, 45.
— 1927 (b): The H. i. c. of plant tissues. VI. Stem tissue reactions throughout the year. Protoplasma **2**, 3, 428.
Rediske, J. H., and G. Biddulph, 1953: The absorption and translocation of iron. Plant Physiology **28**, 4, 576–593.
Reiss, P., 1926: Le pH intérieur cellulaire. Paris.
Ricci, J. E., 1952: Hydrogen Ion Concentration. Princeton.
Richards, T. W., 1898: The relation of the taste of acids to their degree of dissociation. Am. Chem. J. **20**, 121.
Rietsema, J., 1950: Action and penetration of growth substances with special reference to *Avena* coleoptile sections. Meded. Bot. Lab. Utrecht 1950, No. 1.
Rijven, A. H. G. C., 1952: In vitro studies on the embryo of *Capsella bursapastoris.* Meded. Bot. Lab. Utrecht 1952, No. 1.
Roberts, O., and J. Doyle, 1938: The pH of conifer leaves in relation to systematy. Sci. Proc. Roy. Dubl. Soc. **21**, 655–674.
Rogers, C. H., and J. W. Shive, 1932: Factors affecting the distribution of iron in plants. Plant Physiology **7**, 227.
Rohde, K., 1917: Untersuchungen über den Einfluß des freien H-Ions auf den Vorgang der vitalen Färbung. Pflügers Archiv **168**, 411.
Sachs, J., 1862: Über saure, alkalische und neutrale Reaktion der Säfte lebender Pflanzenzellen. Bot. Zeitung **20**, 264.
Sakamura, T., 1927: Chromosomenforschung an frischem Material. Protoplasma **1**, 537.
Samuel, G., 1927: On the shot-hole disease caused by *Clasterosporium carpophilum* and on the "Shot Hole" Effect. Ann. Bot. **41**, 375.
Sayre, J. D., 1926: Physiology of Stomata of *Rumex patientia.* Ohio Journ. of Science **26**, 233.
Scarth, G. W., 1926: The influence of h. i. c. on the turgor and movement of plant cells with special reference to stomatal behaviour. Proc. Internat. Congress Plant Sci., Ithaca. Publ. 1929, 1151–1162.
— 1932: Mechanism of the action of light and other factors on stomatal movement. Plant Physiology **7**, 481.
— J. Whyte and A. Brown, 1933: On the cause of night opening of stomata. Trans. Roy. Soc. Canada **5**, 115.

Scarth, G. W., and M. Shaw 1951 (a): Stomatal movement and photosynthesis in *Pelargonium*. I. Effect of light and carbon dioxide. Plant Physiology **26**, 207.
— 1951 (b): II. Effects of water deficit and of chloroform: photosynthesis in guard cells. Plant Physiology **26**, 581.

Schaede, R., 1924: Über die Reaktion des lebenden Plasmas. Ber. deutsch. Bot. Ges. **42**, 219.

Shive, J. W., and R. A. Ingalls, 1931: Relation of H-ion concentration of tissue fluids to the distribution of iron in plants. Plant Physiology **6**, 103–125.

Schley, E. O., 1913: Chemical and physical changes in geotropic stimulation and response. Bot. Gaz. **56**, 480.
— 1920: Geo-presentation and Geo-reaction. Bot. Gaz. **70**, 69.

Schwarz, F., 1892: Die morphologische und chemische Zusammensetzung des Protoplasmas. Beitr. Biol. Pflanzen **5**.

Scott-Moncrieff, R., 1932: A note on the anthocyanin pigments of the primrose *Primula acaulis*. Jour. Genetics **25**, 206.
— 1936: A biochemical survey of some Mendelian factors for flower colour. Jour. Genetics **32**, 117–170.

Seifriz, W., and M. Zetzmann, 1935: A slime mould pigment as indicator of acidity. Protoplasma **23**, 175–179.

Seifriz, W., 1936: Protoplasm. New York.

Sideris, C. P., and H. Y. Young, 1944: Effects of iron on chlorophyllous pigments, ascorbic acid. acidity and carbohydrates of *Ananas comosus* (L.) Merr. Plant Physiology **19**, 52–75.

Sinke, N., 1937: An experimental study on the structure of living nuclei in the resting stage. Cytologia, Fujii Jub. Vol.
— 1939: Experimental studies of cell-nuclei. Mem. Coll. Sci., Kyoto Imp. Univ. B, **15**, No. 1, 1–116.

Small, J., 1918: Changes of electrical conductivity under geotropic stimulation. Proc. Roy. Soc. Lond., B. **90**, 349.
— 1920: A theory of geotropism. New Phytologist **19**, 49.
— 1923: Propagation by cuttings in acidic media. Gardeners' Chronicle **73**, 244.
— 1926: H. i. c. of plant tissues. I. The method. Protoplasma **1**, 324–333.

Small, J., S. H. Martin and M. W. Rea, 1926: The reaction of plant tissues. Rep. Brit. Assoc., Oxford, p. 410.

Small, J., and M. W. Rea, 1926–27: H. i. c. of plant tissues. II, V, VI. Protoplasma **1** and **2** (see Rea).

Small, J., and C. T. Ingold, 1928: H. i. c. of plant tissues. IX. Improved technique for the R. I. M. Protoplasma **3**, 458.

Small, J., 1929: Hydrogen-ion concentration of plant cells and tissues. Protoplasma Monographs, No. 2. Berlin.

Small, J., and K. M. Maxwell, 1939: pH phenomena in relation to stomatal opening. I. *Coffea arabica* and some other species. Protoplasma **32**, 272.

Small, J., M. I. Clarke and J. Crosbie-Baird, 1942: pH phenomena in relation to stomatal opening. II.—V. Proc. Roy. Soc. Edin. **61**, 233–266.

Small, J., 1946: pH and Plants. London and New York.

Small, J., and T. Jackson, 1949 (a): Relative buffer-index values for root saps of some crop plants. Journ. Agric. Sci. **38**, 343–344.
— — 1949 (b): Buffer index values in relation to soil-pH tolerances. Plant Physiology **24**, 75–83.

Small, J., 1950: A note on sugar and starch in stomatal guard-cells. New Phytol. **49**, 274–276.
— 1952: The new outlook on pH. Protoplasma **41**, 273–276.
— 1954: Modern aspects of pH. London.
— 1955: Estimation of pH values. Subsection F of Section II. General methods, in "Modern methods of Plant Analysis." Springer, Berlin (in the press).

Smith, E. P., 1923: Effects of anaesthetics on plants. Nature **112**, 654.
— 1924: The effect of general anaesthetics on the respiration of cereals. Ann. Bot. **38**, 261.
— 1928: The reaction of the medium in relation to root formation in *Coleus*. Trans. Bot. Soc. Edin. **30**, 53–58.
— 1931: Flower colours as natural indicators. Bot. Soc. Edin. **30**, 230–238.
— 1933: The calibration of flower colour indicators. Protoplasma **18**, 112–124.

SÖRENSEN, S. P. H., 1909: Études enzymatiques. II. Sur la mesure et l'importance de la concentration des ions hydrogène dans les réactions enzymatiques. Compt. Rend. Lab. Carlsberg **8**, 1 (note definition of pH on p. 28, not that on p. 4).

SPEK, J., and R. CHAMBERS, 1933: Das Problem der Reaktion des Protoplasmas. Protoplasma **20**, 376–406.

STANFIELD, J. F., 1937: Certain physico-chemical aspects of sexual differentiation in *Lychnis dioica*. Amer. J. Bot. **24**, 710–719.

— 1944: Chemical composition of roots and tops of dioecious *Lychnis* in vegetative and flowering phases of growth. Plant Physiol. **19**, 377–383.

THODAY, D., and H. EVANS, 1931: Buffer systems in *Kleinia articulata*. Protoplasma **14**, 64–74.

— — 1932: The distribution of calcium and phosphate in the tissues of *Kleinia articulata* and some other plants. Ann. Bot. **46**, 781–806.

— — 1933: The distribution of some solutes in the tissues of *Kleinia articulata*. Ann. Bot. **47**, 1–20.

THODAY, D., and M. W. P. JACKSON, 1939 (a): The distribution of calcium malate and other solutes in the stems and leaves of succulent Compositae. Ann. Bot. N. S. **3**, 1–26.

THODAY, D., and K. M. JONES, 1939 (b): Malic acid and respiration during starvation in *Kleinia articulata*. Ann. Bot. N. S. **3**, 677–698.

THOMAS, M., 1951 (a): Vegetable acids in higher plants. Endeavour **10**, 160–165.

— 1951 (b): Carbon dioxide fixation and acid synthesis in Crassulacean acid metabolism. Symposia of Soc. for Exp. Biology **5**, 72–93.

TOMICEK, O., and J. FELDMAN, 1934: The determination of acidity in non-aqueous solutions. Coll. Czechoslov. Chem. Comm. **6**. See Chem. Abst. 1935, **29**, 704 for oils and solutions: 6171 for more solutions.

TRUOG, E., 1918: Soil acidity. I. Its relation to the growth of plants. Soil Science **5**, 169–195.

— and M. R. MEACHAM, 1919: Soil acidity. II. Its relation to the acidity of the plant juice. Soil Science **7**, 469.

ÚLEHLA V., and V. MORAVEK, 1922: Über die Wirkung von Säuren und Salzen auf *Basidiobolus ranarum* Eid. Ber. deutsch. bot. Ges. **40**, 9.

ÚLEHLA, V., 1928: Geweberegulation bei Sukkulenten. Protoplasma **3**, 469.

VAN SANTEN, A. M. A., 1938: Influence of hydrogen-ion concentration on the growth rate of the *Avena* coleoptile. Proc. Konin. Neder. Akad. v. Weten. **41**, 513–523.

— 1940: Groei, Groeistof en pH. Thesis. Utrecht.

VAN SLYKE, D. D., 1922: On the measurements of buffer values and on the relationship of buffer-value to the dissociation constant of the buffer and the concentration and reaction of the buffer solution. Jour. Biol. Chem. **52**, 525.

VICKERY, H. B., 1952: The behaviour of isocitric acid in excised leaves of *Bryophyllum calycinum* during culture in alternating light and darkness. Plant Physiology **27**, 9.

WALBUM, L. E., 1913: Sur l'emploi de l'extrait de choux rouge comme indicateur dans la mesure colorimétrique de la conc. d. ions hydrogène. C. R. Lab. Carlsberg **10**, 227.

WALLING, C., 1950: The acid strength of surfaces. J. Amer. Chem. Soc. **72**, 1164–1168.

WATSON, G. N., 1913: The juice of the blueberry as an indicator. Am. J. Pharm. **85**, 246.

WEBER, F., 1923 (a): Zur Physiologie der Spaltöffnungsbewegung. Österr. botan. Zeitschr. 1923, No. **1–3**, 43–57.

— 1923 (b): Enzymatische Regulation der Spaltöffnungsbewegung. Die Naturwiss., Wien, 2, **17**, 309–316.

— 1926: Die Schließzellen. Archiv f. exper. Zellforschung **3**, 101.

— 1927: Vitale Blattinfiltration. Protoplasma **1**, 581–588.

— 1930: Vakuolen-Kontraktion, Tropfenbildung und Aggregation in Stomata-Zellen. Protoplasma **9**, 128–132.

WILLIAMS, W. T., and M. E. SHIPTON, 1950: Stomatal behaviour in buffer solutions. Physiologia Plantarum **3**, 479–486.

— 1952: Studies in Stomatal Behaviour. II. The rôle of starch in the light response of stomata. J. exper. Bot. **3**, 110–127.

WILLSTÄTTER, R., 1914: Farbstoffe der Kornblume. Liebigs Annalen **401**, 189.

WOOD, W. M. L., 1952: Organic acid metabolism of *Sedum praealtum*. Journ. Exper. Bot. **3**, 9, 336–355.

YAMAHA, G., and T. ISHII, 1932: Über die Ionenwirkung auf die Chromosomen der Pollenmutterzellen von *Tradescantia reflexa*. Cytologia **3**.
— 1933: Über die Wasserstoffionkonzentration und die isoelektrische Reaktion der pflanzlichen Protoplasten, insbesondere des Zellkernes und der Plastiden. Protoplasma **19**, 193–212.
ZIRKLE, R. E., and W. BLOOM, 1953: Irradiation of parts of individual cells. Science **117**, 493–496.

Protoplasmatologia
II. Cytoplasma
B. Chemie
2. Cytochemie und Histochemie
c) The pH of Animal Cells

The pH of Animal Cells

By

FLOYD J. WIERCINSKI

Hahnemann Medical College, Dept. of Physiology, Philadelphia, Pa.

With 7 Figures

Contents

I. Introduction

From the physicochemical point of view protoplasm has been defined as a colloidal solution of amphoteric electrolytes. To understand the behavior of such a solution it is necessary to know the hydrogen ion concentration. This, beyond doubt, is one of the most important variables. Changes in hydrogen ion concentration can have serious effects on the behavior of protoplasm; for example, on enzymatic activity, and on the state of aggregation of proteins.

Numerous attempts have been made to determine the pH of various parts of different cells and organs. The results obtained by these investigations are not always in agreement. The early literature is reviewed in a monograph by REISS (1926 a). Later, GRÄFF and RAPPOPORT (1937) outlined methods and results. LISON (1941) compiled data on intracellular pH and rH. For references to more recent papers CHAMBERS (1949) and HEILBRUNN (1952) may be consulted. The picture presented in the literature, as a whole, is not a consistent one. A half century of research has not given the answer to the problem of intracellular pH.

In this discussion an attempt will be made to gather and evaluate all the existing literature and data on pH in the protoplasm of animal cells. It is intended to present to the interested investigator the problems, methods and results obtained to the present time. The data are tabulated for each phylum from the Protozoa through the Chordata. Methods are carefully examined in order to evaluate their sources of error, and the assumptions upon which their procedures are based. Some of the contradictory values in the literature are certainly due to variations in technique and errors in the methods used.

A review of the literature on pH in animal cells is rather discouraging. Values obtained for a given cell may differ by as much as two units of pH. A large part of the variation is undoubtedly due to lack of precision in the method used. In order to know the true pH of any part of a given cell it would be necessary to apply a coefficient of correction for salt, protein, and errors involved in observation (LISON 1941).

The manner in which the living cell is organized is certainly one of the fundamental unsolved problems of all biology. In the past, cell physiologists have applied the concepts of physical chemistry to the study of the cell. At times, these studies have introduced new ideas and concepts into the science of physical chemistry. It is hoped that this review will serve not only to show the present status of our knowledge of the hydrogen ion concentration of protoplasm but may also stimulate a rather lagging interest in a field of great importance to cellular physiology.

II. Definition of pH

Living protoplasm is not a homogenous system and it must be remembered that the living cell contains various constituents such as a nucleus, nucleolus, granules, fat droplets, and vacuoles. We have to consider the pH of the hyaline protoplasm and, perhaps, also the pH of the nucleus and of various protoplasmic inclusions.

With the work of Arrhenius (1887 a, b, 1889) came the concept of acidity resolved into two components. One is "normality," which, in the older sense, is the quantitative factor that describes the total acid content, while the other is the hydrogen ion concentration, which is the intensity factor. The expression of hydrogen ion concentration, namely, pH, was introduced by Sörenson (1909 a, b).

It is important that the biologist should have a good working concept of the definition of pH and that he should be aware of factors in living cells that affect pH. The definition of Sörensen (1909 a, b) in terms of hydrogen ion activity is accepted to this day. Sörensen defined pH (Glasstone 1946, p. 982) as

$$pH = \frac{1}{\log a_{H^+}}$$

It follows that

$$pOH = \frac{1}{\log a_{OH^-}}$$

Since it is difficult to measure the activity of single ions, the approximate expression

$$pH = \frac{1}{\log C_{H^+}}$$

has come to be the accepted definition.

Many details have been introduced by other workers to fit conditions of their experiments, and equations have been modified accordingly but no one has improved upon the basic definition.

No attempt will be made to outline the physical chemistry of hydrogen ion concentration measurement. There are excellent chapters in Glasstone (1946) and Reilly and Rae (1953) on the theory and practice of pH determination in aqueous solutions. Ricci (1952) devotes an entire book to the mathematical treatment of hydrogen ion concentration. Small (1955) quotes the fundamental equilibrium constant from Ricci (1952).

From the point of view of the application of hydrogen ion concentration to biological processes Ogston (1947) has written on the definition and meaning of pH. Albert (1952) discusses ionization, pH and biological activity. Small (1952, 1955) urges caution in evaluating the significance of hydrogen ion concentration and the accuracy of the procedures. pH cannot be regarded as a measure of hydrogen ion concentration, but as a measure of hydrogen ion activity, which in turn reflects the activity of associated ions and of the solvent. It must be remembered that Sörensen's original definition was in terms of hydrogen ion activity. As a result of this, Small offers a new definition of pH as a "series of values for the differences between the unknown and one or more solutions of 'standard' pH values."

III. The Problem of pH in Protoplasm

The determination of pH in the living cell is much more difficult than the determination of pH in an aqueous solution. The cell has structural organization. It is composed of a cell wall, nucleus and cytoplasm in which there are contained vacuoles and granules.

SCARTH (1924) raised the question of whether the hydrogen ion concentration of living protoplasm could be determined. In his work with the large ameba, *Pelomyxa*, SCARTH concluded that the liquid of the vacuoles has a different reaction from the cytoplasm and that the nuclear reaction is presumably quite different.

LUCKÉ (1925) centrifuged *Arbacia* eggs after staining with neutral red and brilliant cresol purple. The cytoplasm remained colorless and the dyes were in the cytoplasmic inclusions. SPEK (1934 b) used vital dyes on *Asterias* eggs and found the pH to vary from 5.0 to 7.6 for various cytoplasmic inclusions. These studies indicate that the hyaline protoplasm is unaffected by certain dyes. Studies on pH in the hyaline protoplasm were made by WIERCINSKI (1944). He centrifuged cells and then microinjected them with sulfonphthalein indicators.

PETERS (1929) has calculated from the equation $C_H X C_{OH} = k_w$ that there is one molecule dissociated in some 555,000,000 molecules of water at ordinary temperature. The volume occupied by this amount of water is a cube with a side of 0.25 μ length.

GORTNER (1930) states that water may exist in two states: (1) as free water which is characteristic of water in bulk, and (2) bound water which is characteristic of the lyophilic system. The equilibrium between free and bound water is of major importance in vital phenomena.

BERNAL and FOWLER (1933) proposed to change the name hydrogen ion (hydrion) to oxonium ion, which could be represented by the symbol OH_3^+. This symbol is also written as H_3O^+ and is also called the hydronium ion. Water is dissociated according to the following equation

$$2\,H_2O \rightleftharpoons H_3O^+ + OH^-$$

Also, cells are composed of numerous surfaces, DANIELLI (1941, 1946) and HARTLEY and ROE (1940) claim that the pH of a surface may differ from the fluid which bathes it by two or more pH units.

The nucleus, being more or less in the center of the cell, presents difficulties from the point of view of intracellular pH determination by means of indicator dyes and microelectrode technique.

CHAMBERS (1943) discussed the problem of solutions which exert no damage to intracellular components. He devised a solution for *Physarum* plasmodium that was found to be compatible for the interior of fresh water ameba. The composition of this solution was KCl 0.120 M, NaCl 0.013 M, and $CaCl_2$ 0.003 M in water. The ratio of K^+ to Na^+ was 9 and that of K^+, Na^+ to Ca^{++} was 41. It is interesting to note that *Physarum* can withstand a high calcium concentration whereas in the *Echinoderm* egg even a trace of calcium will cause a breakdown of intracellular granules. There is the possibility that a suitable intracellular solution should contain some organic material.

IV. Methods Used for the Determination of Intracellular and Tissue pH

For the determination of pH in protoplasm the method employed is usually a static one. The preferred method would be one that would have a minimum of interference of a known nature.

The methods for the determination of intracellular pH are of three distinct types, as follows: (A) potentiometric method, (B) indicator method and (C) methods of calculation: (1) involving the use of the Henderson-Hasselbalch equation, (2) the Zeta potential and (3) buffering power.

A. Potentiometric Methods

With the potentiometric method a pH meter is used to measure the electromotive force. Extracts of cells and tissues have been made by freezing, grinding, and even boiling the extracts to drive off the carbon dioxide when a determination was made. Various types of electrodes such as the antimony, hydrogen and microelectrodes of glass, platinum and tungsten have been inserted into cells. Glass electrodes have also been placed in contact with tissues to make pH determinations.

1. Hydrogen, platinum and antimony electrodes

MICHAELIS and KRAMSZTYK (1914) made electrical determinations with the hydrogen electrode on extracts of mammalian tissues. They boiled the extracts to destroy enzymes so that carbon dioxide would not be produced as a result of enzymatic action.

BODINE and FINK (1925) constructed a simple microvessel with an electrode for determining the hydrogen ion concentration of small amounts of fluid. They applied their technique to the study of body fluids.

KERRIDGE and WINTON (1929) minced cat and rabbit uterine muscle and found that it was very little affected by the solution in which it had been immersed. Measurements of the minced tissue indicate that the hydrogen ion concentration of the isolated uterus is greater than that of the uterus in situ.

BUYTENDIJK and WOERDEMANN (1927) used the micro-antimony electrode to measure the pH of amphibian eggs; they inserted the electrodes into the eggs. DORFMAN (1938) and DORFMAN and GRODENSKY (1937) used the micro-platinum hydrogen electrode in developing amphibian eggs, inserting the electrodes into the egg interior. This work has a source of error in that it does not take into account the possibility of a membrane potential around the inserted electrode. The results obtained as seen in the tables under *Results* are far from being in agreement. Also, the fact is inescapable that the electrode was inserted into the yolk substance.

The measurement of pH of the cell with the potentiometric method at the present time is both difficult and uncertain. The procedure is open to criticism because the micro-electrode causes a local cytolysis and only measures the pH of the injured region.

VLÈS, REISS, and VELLINGER (1924 a, b and c) froze and ground up sea urchin eggs and measured the pH with the hydrogen electrode.

VELLINGER (1927) made a similar study by grinding the eggs in sea water into a paste and making a pH determination of the paste at -6^0 C. There were two sources of error. There was contamination of the sea urchin paste with sea water which has a pH of 8.2 and also, a temperature correction was necessary. The interesting point about these determinations is that the methods used have yielded rather acid pH values for the cells examined. See Fig. 1 for a diagram of the mortar used by REISS and VELLINGER.

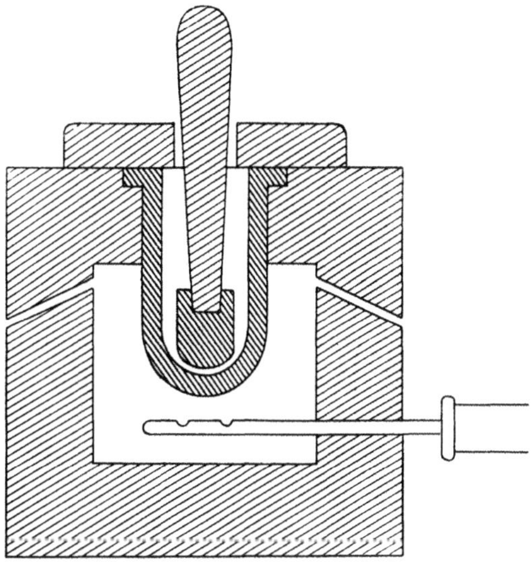

Fig. 1. Model of mortar used by REISS and VELLINGER (1926) to hold liquid carbon dioxide for the grinding of tissues at a low temperature. Redrawn from REISS (1926 a).

VELLINGER (1928) calibrated the antimony electrode against the hydrogen electrode to determine the salt and protein errors. The salt error in 0.25 M phosphate plus sucrose from 0.0 to 0.75 M concentration varied from -0.04 to $+0.42$ pH units. 0.05 M potassium phthalate plus KCl from 0.0 to 1.5 M concentration showed a salt error from -0.05 to $+0.65$ pH. The protein error was determined on egg white, casein, and gelatin. The protein error for various concentrations of protein varied from -0.09 to $+0.08$ pH.

2. Capillary glass electrode

For the determination of pH in tissues under conditions of inflammation SCHADE, NEUKIRCH and HALPERT (1921) designed a subcutaneous electrode.

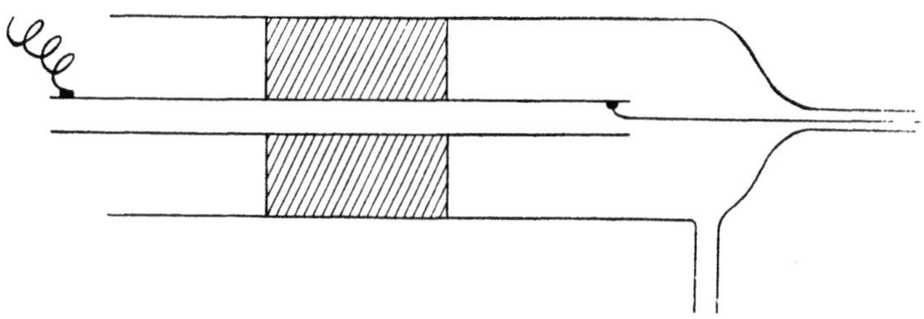

Fig. 2. Subcutaneous electrode SCHADE, NEUKIRCH and HALPERT (1921).

See Fig. 2 for a diagram of the electrode. This was used in human tissues. Also, see NIMS (1938).

3. Glass electrode

Millet (1928) used the glass electrode on tissues frozen in liquid air and the pH was then determined at 18° C. The tissues used were rat carcinomas and human neoplasms. The pH obtained by this method indicated a lower value than that of the surrounding normal tissues.

Dubuisson (1936, 1939) found pH changes of muscle during and after contraction. The principle used was to cover a part of the surface of the muscle with a glass electrode and to record the pH changes occurring on stimulation. The thin liquid film outside the muscle, which is in contact with the glass membrane, begins to become more alkaline in the course of the contraction.

This method permits the study, without any damage to the muscle fibers, of the surprisingly fast exchange of CO_2 which results from chemical processes occurring during and after contraction between the inside of the muscle and the moist film covering its surface.

Dubuisson (1939) used a Corning glass electrode to measure the alterations in pH which occur at the surface of a muscle in the course of its activity. The work was done on frog and tortoise smooth muscle. Four seconds were required to record a deflection on photographic paper for an instantaneous deflection in pH under the glass membrane electrode. See Dole (1941) for a discussion of the glass electrode. Dubuisson (1950) has also used the glass electrode to follow pH changes at the surface of frog muscle during contraction and relaxation.

In the course of the isometric contraction of smooth muscle, four successive changes in reaction were produced, designated by the letters (a), (b), (c) and (d).

(a) An alkaline phase, beginning before the contraction, maximum at the outset of the period of increasing tension. This corresponds to an unknown chemical process.

(b) An acid phase. Begins before contraction, maximum toward the middle of the period of rising tension. It is proportional to the tension developed. It is probably an expression of the decomposition and resynthesis of adenyl pyrophosphoric acid.

(c) An alkaline phase. Begins during the period of rising tension, maximum at the outset of relaxation. It represents the hydrolysis and synthesis of creatine phosphoric acid.

(d) An acid phase. Begins during contraction, maximum several minutes later. It is an expression of the formation of lactic acid.

Disteché and Dubuisson (1954) point out that previous techniques were too slow to study the pH changes during a single twitch of a striated muscle. They offer a direct current amplifier with a electrometer input stage which makes it possible to follow pH changes on the oscilloscope screen. The sensitivity is 0.003 pH unit per centimeter of beam deflection. The response is independent of glass membrane thickness from 5 to 10 μ. There are difficulties with unclean glass membranes. The pH change is slowed considerably.

CALDWELL (1953, 1954) used micro-glass and micro-tungsten electrodes in the 500 to 600 micra muscle fibers of *Carcinus maenas* (Crustacea). He found the intracellular pH to be about 6.9. The micro-glass electrodes are more suitable. The micro-tungsten electrodes appear to be affected by many other factors. See CALDWELL and HARRIS (1952) for a statement on metabolic control by intracellular pH.

B. Indicator Method

1. General considerations

The method involves matching colors of a test solution with a buffer standard at a known pH. With aqueous solutions the conditions can be controlled very precisely but with living cells many problems arise. The cell is not a homogenous solution, it has a membrane in which the contents are held together. The interior is composed of a liquid and a solid phase in the form of hyaloplasm, granules, vacuoles, inclusion bodies and a nucleus. Vital dyes have been used with living cells simply by immersing the cell in a solution of the dye. The cellular constituents adsorb these dyes and are a source of error. The best indicators for the determination of pH by colorimetric means are the acid-base indicators. These, however, do not stain cells very easily. The methods that have been used to introduce them into the living cell have involved crushing the cell in indicator solutions or microinjection.

a) Colorimetric pH determination and definition

KILPATRICK (1935) discusses the correct definition of pH in relation to the colorimetric determination of hydrogen ion concentration. In the use of the colorimetric method great care is often taken in making the measurements and all too little thought is given to the calculation of the result. This is indeed the case with the determination of intracellular pH. All the work has been directed to the observation of a direct reading of a pH value in protoplasm but rarely has any attempt been made to evaluate sources of error.

The substances used for the colorimetric determination of hydrogen ion concentration are acids and bases, of which the acid form or basic form or both are colored, KILPATRICK and KILPATRICK (1932) and KILPATRICK (1935). In the case of a weak monobasic acid, HI, the double acid-base equilibrium with the solvent can be represented in the following manner.

$$\underset{\text{acid}}{HI} + \underset{\text{base}}{H_2O} - \underset{\text{acid}}{H_3O^+} + \underset{\text{base}}{I^-} \qquad (1)$$

By increasing the concentration of the base I^-, the solution can be buffered in respect to the hydrogen ion H_3O^+ (also known as hydronium and oxonium ion). If HI or I^- is colored we have an indicator. The fact that the indicator establishes its own buffer equilibrium is of great importance in the determination of the hydrogen ion concentration of unbuffered or poorly buffered solutions. For example, if we are determining the hydrogen ion concentration of carbon dioxide-free water, with a hydrogen ion concentration of I X 10^{-7} M, and we add an indicator, the hydrogen ion concentration is no longer I X 10^{-7} M. In the case of well buffered solutions this problem does not arise, for here the constituents of the buffer system of the indicator are present at concentrations of the order of 10^{-5} M, while the constituents of the buffer system whose hydrogen ion concentration we are measuring are usually present at concentrations of the order of 10^{-1} or 10^{-2} M.

The equilibrium involved can be formulated by the mass law as follows:

$$K_c = \frac{C_{H_3O} + C_{I}^{-}}{C_{HI}}$$

where K_c represents the classical dissocation constant. When the concentration of HI is equal to that of I^-, that is when the indicator is half transformed, we have:

$$K_c = C_{H_3O^+}$$

Now K_c varies with the electrolyte concentration so that the half change point in one solution may not be the half change point in another, even though the hydrogen ion concentrations are the same. This means that equal color for equal indicator concentration does not always show equal hydrogen ion concentration. Also, it does not show equal hydrogen ion activity. The fundamental equation for two solutions of equal indicator concentration and equal color is

$$\left(\frac{C_{H_3O^+}}{K_c}\right) I = \left(\frac{C_{H_3O^+}}{K_c}\right) II$$

Letting solution I be the known, there results for the hydrogen ion concentration of the unknown solution

$$(C_{H_3O^+})\, II = \frac{(C_{H_3O^+})\, I\ (K_c)\, II}{(K_c)\, I}$$

If the ratio $(K_c)\, II/(K_c)\, I$ is known as well as $(C_{H_3O^+})\, I$, $(C_{H_3O^+})\, II$ can be calculated.

b) Hydrogen ion indicator chart

Chart of indicators and their ranges of pH. Reprinted by permission from the National Aniline Division of the Allied Chemical and Dye Corporation, N. Y. (See page 11.)

c) Structural formulae of dyes and indicators

In the study of intracellular pH we are concerned with organic compounds of various types. A typical member of each group is given together with its structural formula as outlined by TOMICEK (1951). Also, see CLARK (1928) and BROCKMAN (1935).

HYDROGEN ION INDICATOR CHART

RANGES AND COLOR CHANGES OF NATIONAL ANILINE DIVISION'S ORGANIC CHEMICAL INDICATORS

The Abbreviations used are as follows

A—Amber	C—Colorless	Pu—Purple
B—Blue	Fl—Fluorescent	R—Red
BG—Blue Green	G—Green	V—Violet
Br—Brown	O—Orange	Y—Yellow
BV—Blue Violet	P—Pink	

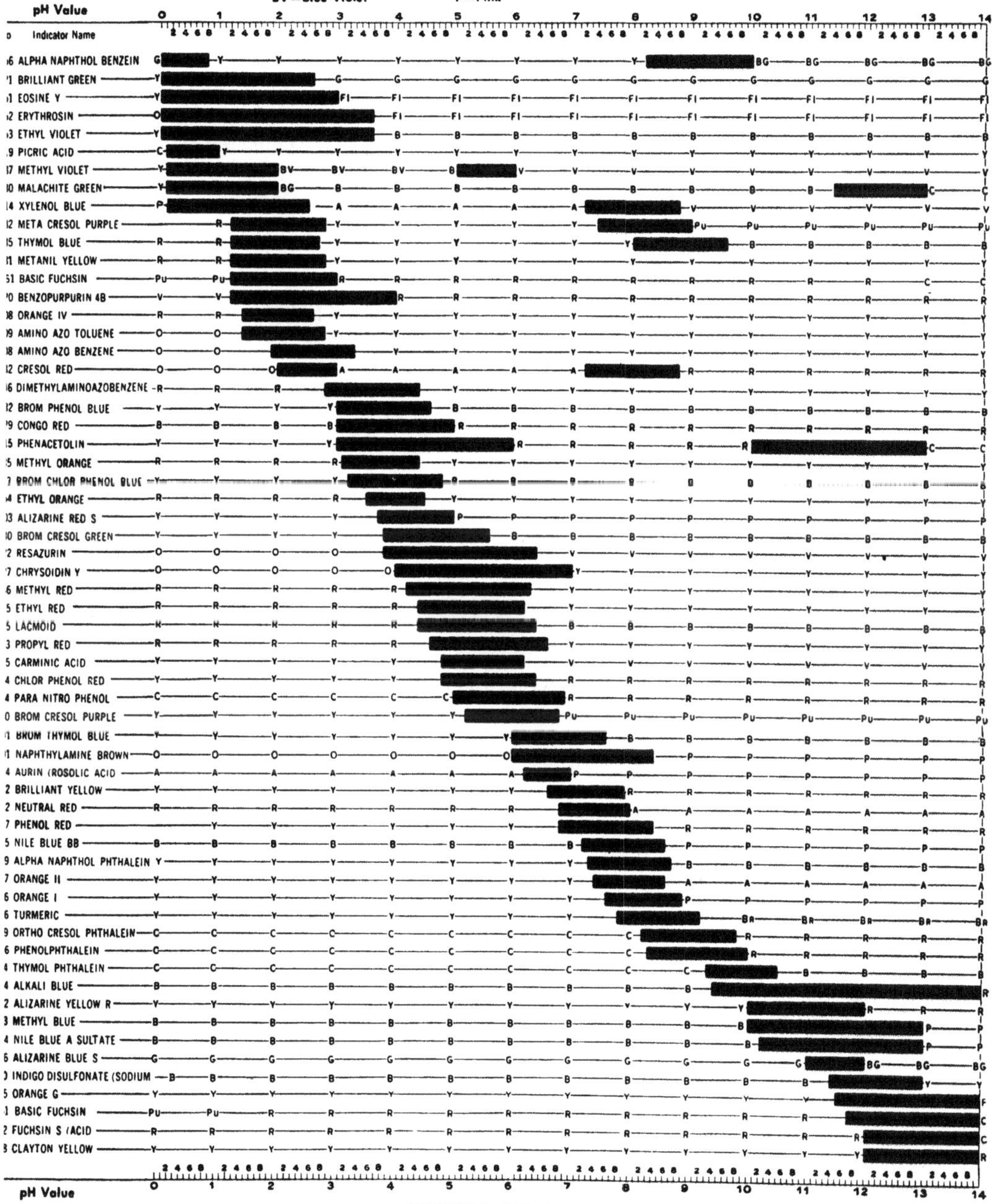

SUGGESTED REFERENCES

Indicators	I M Kolthoff and N Howell Furman
Hydrogen Ions	Hubert T S Britton
The Determination of Hydrogen Ions	W Mansfield Clark

Nitro compounds

p-nitrophenol

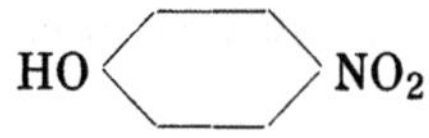

Nitramine 2, 4, 6-trinitrophenylmethyl nitramine

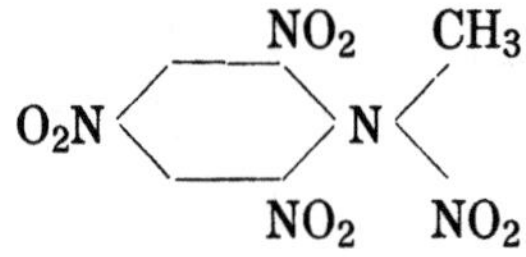

Azo compounds

Methyl orange = sodium dimethylamino-azo-benzene-p-sulfonate

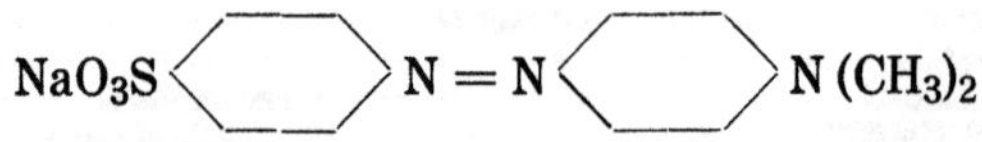

Methyl red = dimethylaminoazobenzene-o-carboxylic acid

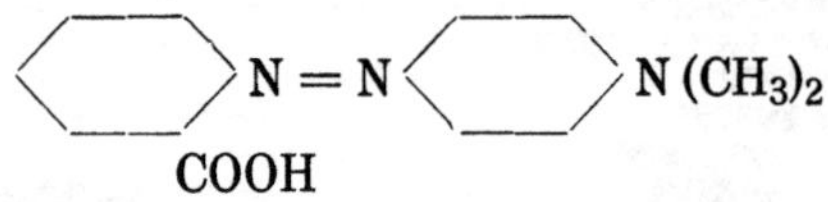

Bisazo compounds

Congo red = diphenyl-4, 4'-bis (azo-2-naphthylamine-1-sulfonic acid-4)

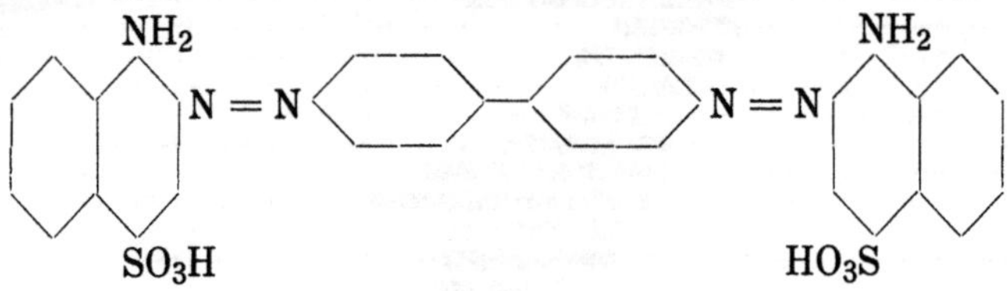

Triphenyl methane dyes

Crystal violet = hexamethyl-p-rosanaline chloride

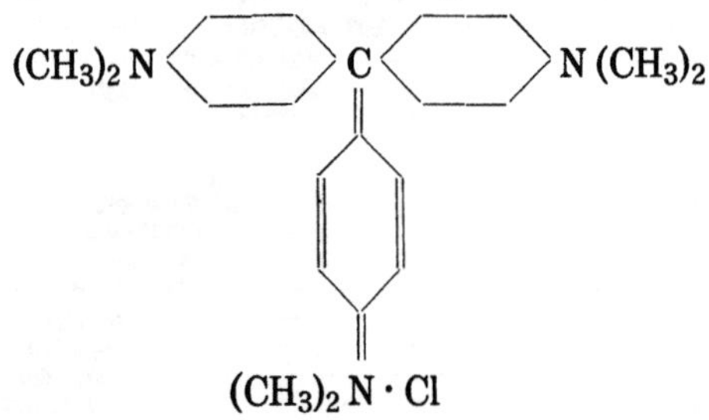

Phthaleins

Phenolphthalein = di-p-dioxydiphenylphthalide

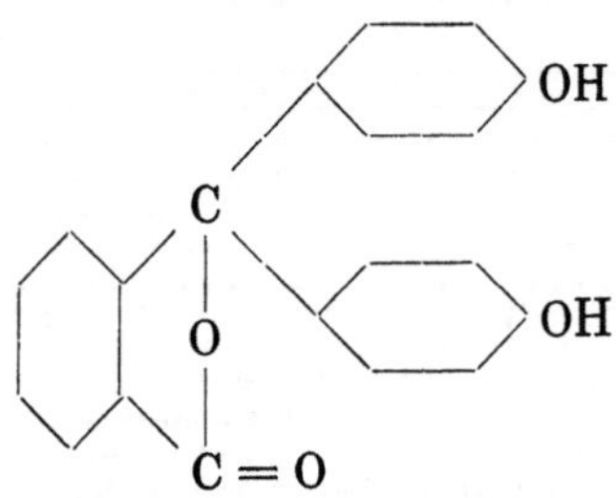

Sulfonphthaleins

Phenol red = Phenolsulphonphthalein

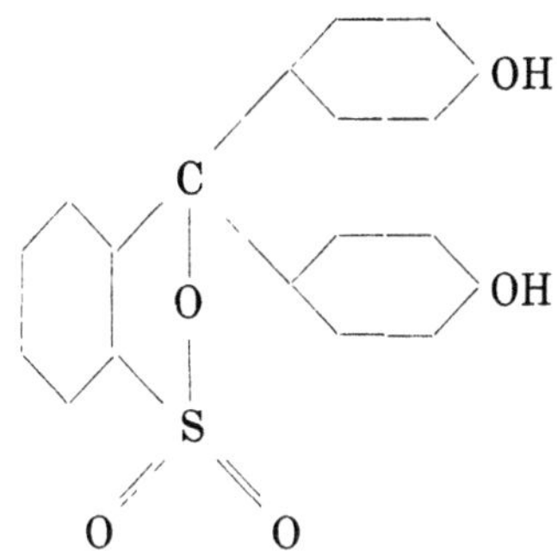

Quinoline dyes

Cyanin = 1, 1′-di-isoamyl-4, 4′-quinocyanin iodide

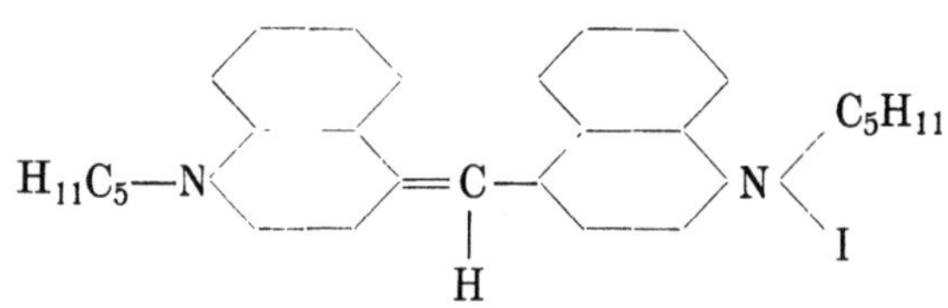

Indophenols

Indophenol = p-oxyphenyl-quinone monoimine

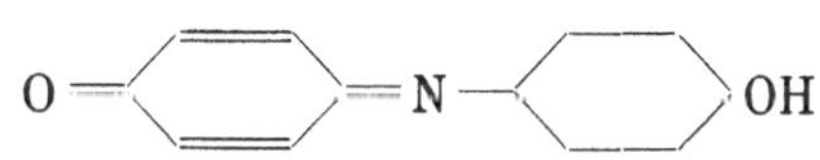

Azine dyes

Neutral red = aminodimethylaminomethylphenazinium chloride

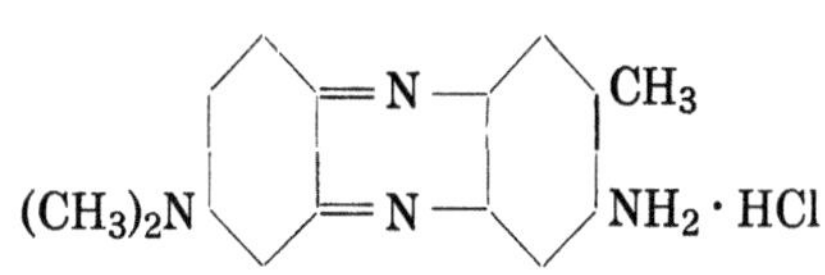

Oxazine dyes

Alizarin green = dioxydinaphthazoxonium sulphonate

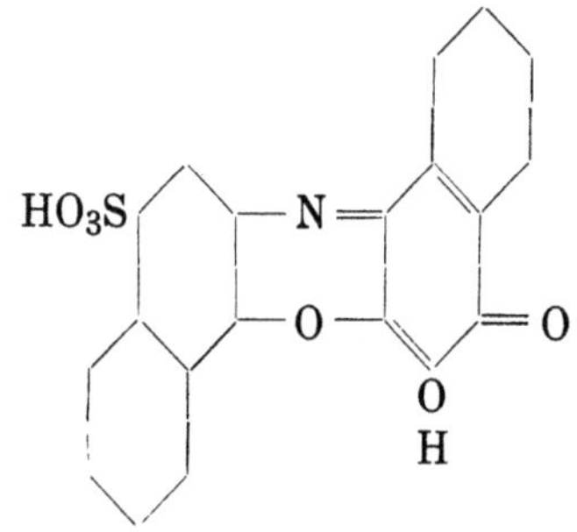

Anthraquinine dyes

Alizarin = 1, 2-dioxyanthraquinone

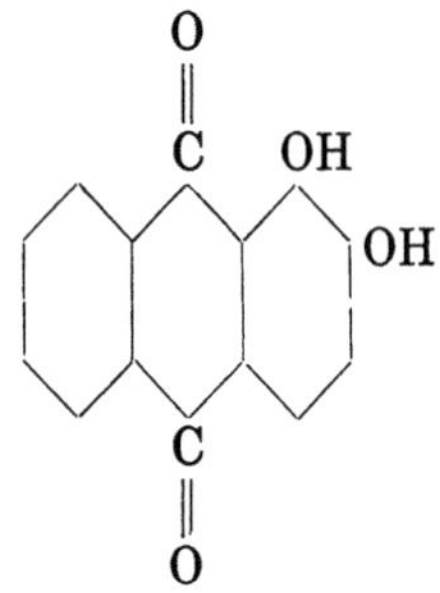

Indigo dyes

Indigo carmine = indigotin-5, 5′-disulphonic acid

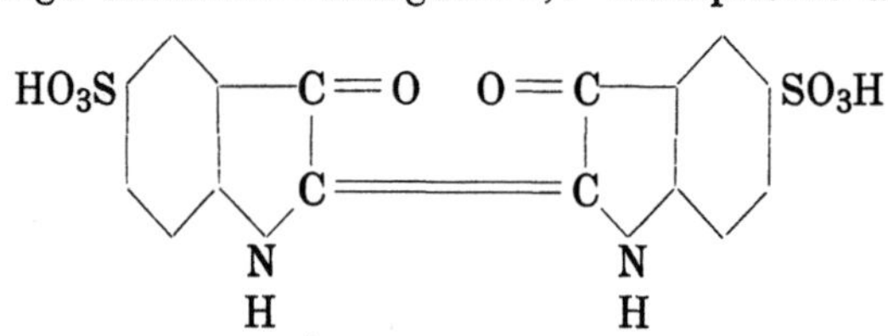

d) The problem of color identification

REISS (1928 a) devised an apparatus for isolating a single egg cell. He verified the law of Beer-Lambert and determined a calibration curve for the dye bromcresol purple when it was injected into the *Paracentrotus lividus* egg cell. The size of the droplet injected into the egg was $^1/_5$ the diameter of the egg. The dilution of the droplet in the egg was $^1/_{125}$. He calculated that 27 to 29% of the alkaline form of the indicator was present. Protein errors were also considered. The pH of the normal unfertilized egg cytoplasm was found to have a value of 5.6. When the egg was injured the pH was 4.7.

SMALL (1955) points out that the observer should be tested for color blindness. SMALL has found that more women than men students have been successful with pH indicator studies.

DANIELS, MATHEWS, and WILLIAMS (1941) and GLASSTONE (1946) discuss Beer's Law in relation to incident light and absorbing medium. TAYLOR (1931) and EASTMAN and ROLLEFSON (1947) discuss Lambert's absorption law. The importance of these phenomena in relation to the transmission of light through a color standard and the comparison with the color observed in a living cell should not be underestimated. Also, see CLARK (1928) for the "range of usefulness" of an indicator.

The following work illustrates some of the difficulties in ascertaining the color of small objects.

WILLMER (1944) stated that if a circle 2 cm. in diameter, divided into quadrants and colored as shown in the diagram, is placed on either a black or white ground, and then viewed with one eye from a distance of about 3 meters, it is found that there is considerable difficulty in distinguishing the green from the light blue on one hand and the orange-brown from the light red-purple on the other, so long as the brightness of the fields are made as equal as possible. There was no difficulty in distinguishing the

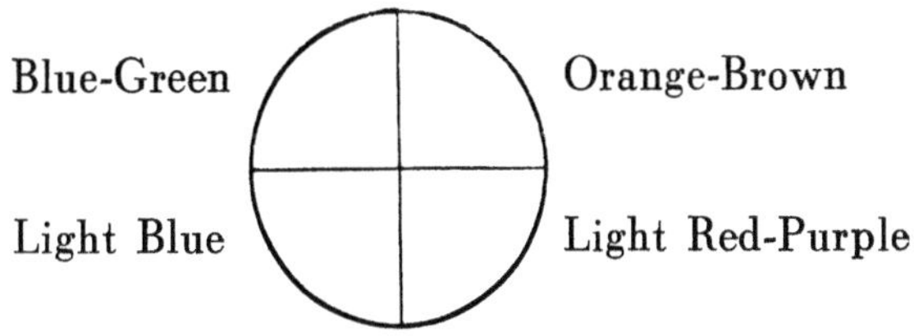

green from the orange, provided that the observer was not red-green color blind. Such observers saw the circle as uniform, or they were able to separate the quadrants on subtle brightness differences.

Follows (1934) studied the mistakes made by persons with defective color vision. It seems that they can be graded in a continuous series showing more or less confusion in the "yellow" of the spectrum extending into the "green" and "red" regions.

Murray and Spence (1939) and Murray (1952) found three kinds of yellow that are indistinguishable to the naked eye. Colorimeters are a help in the duplication and matching of test colors.

e) Matching colors against a standard

In any colorimetric determination of pH it is necessary to prepare a buffer standard. A known volume is taken to which is added a known amount of the indicator or dye. It is not always easy to maintain this relationship when indicators or dyes are microinjected or when the indicators are allowed to diffuse into the living cell.

In the preparations of a buffer indicator standard to be matched against an indicator injected into a living cell the buffer concentration of the living cell should be known so as to minimize error made in the comparison.

The matching of a flash of color occurring in a living cell against a capillary standard in the field of a compound microscope is done very quickly. In order to ascertain the pH, one must observe rapidly and make a decision on the basis of the original flash of color that follows the injection of an indicator; otherwise secondary reactions may occur resulting in false values.

According to Small (1952) a tint comparison is accurate to ± 0.1 pH units.

2. Vital dyes

The simplest technique consists of immersing cells in a vital dye solution, which stains the cellular constitutents. See the diagram for illustration of the apparatus, Fig. 3. The dye changes color as a function of pH. However, a point is reached when too much dye is taken up in the cell and the color values are obscured. Shortly after staining, a comparison can be made microscopically with a standard of known pH. An example of this type of dye is neutral red. However, such dyes will stain the cytoplasmic vacuoles but will not affect the hyaloplasm.

VONWILLER (1921) and ALEXANDROV (1932) have data of 19th century work done with various vital dyes on organisms. LISON (1935) has shown a "metachromatic change" which cannot be distinguished from the "ionic color change." This is due to the presence of very small amounts of esters of sulfuric acid. Thus, one cannot be sure whether the color change is to be attributed to a change in acidity or the presence of these substances. With such dyes as neutral red, brilliant cresol blue, cresol violet, and Nile blue sulfate in M/10,000 concentration the metachromatic error can amount to 6 units of pH.

MICHAELIS (1947) discusses the nature of the interaction of nucleic acids and nuclei with basic dye stuffs.

Vital staining with methyl red

CHAMBERS (1929 a) stated that the permeability of protoplasm to certain substances is due to the fact that these substances penetrate because of the solubility of their associated molecules in the plasma membrane of the living cell. Methyl red (carboxylic orthobenzene azo dimethyl amino) is an amphoteric electrolyte possessing both a carboxyl and an amino group.

Since methyl red is amphoteric and, by virtue of its basic group, should become dissociated in the pH ranges below 5.0. Attempts were made to ascertain if, in spite of this, vital staining would still occur. This was found to be feasible. At a pH of 4.0 and below both the marine ova and the amebae became moribund. In addition the methyl red became increasingly insoluble.

3. Acid-base indicators

Indicators are affected by various influences and these include temperature, salts, and proteins and their decomposition products TOMICEK (1951).

a) Temperature effects

Temperature in general, has a small effect upon the color change of indicators. There is a considerable effect upon the dissociation constant of water. With a rise in temperature the acid coloration of indicators sensitive to bases is moved to the alkaline side. With indicators sensitive to acids the color change is toward the acid side. It is interesting to note that the dissociation constant of sulfonphthalein dyes remains almost unaffected.

REILLY and RAE (1953) state that the temperature effect can be ignored since the error involved is not greater than that of the method. With some indicators it may be large, for example with phenolphthalein it is 0.15 pH units per degree.

b) Salt effects

Salts in solution have different effects on different indicators. There is a small effect on such dyes as methyl red, methyl orange and phenolphthalein. There is a greater effect of salt on the sulfonphthaleins, and a very considerable effect on litmus and Congo red.

Matsuda and Kayama (1949) determined the ionization exponent of bromthymol blue at 38° C. for phosphate mixtures and solutions of sodium chloride. The molarity of the phosphate mixture was 0.133 and 0.308 for sodium chloride. The pk_2 for bromthymol blue was 7.32 for phosphate, and 7.34 for sodium chloride. Both values should agree. The small difference is probably due to experimental error.

c) Effect of proteins and their decomposition products

These substances are amphoteric and since they can combine with both acidic and basic dyestuffs they affect the color change of indicators. The indicators of simple structure are least affected. P-nitrophenol, and phenolphthalein are relatively little affected.

Reiss (1928 a) indicates that the protein error for bromcresol purple for 0.01% concentration of indicator amounts to a + 0.01 pH at a hydrogen electrode pH of 4.84, — 0.16 at a pH of 5.34, and — 0.05 at a pH of 5.60.

Kolthoff (1937) has shown a large "protein error" for methyl red. In solutions containing serum albumin, methyl red gave values 0.79 to 1.1 pH units greater than the electrometric pH values. For other substances such as pepsin, peptone, and gelatin the differences are small (0.03 to 0.09 pH units).

Danielli (1937, 1941) discusses the work of previous investigators. As a general rule, where the indicators and protein have opposite charges, the protein error may be large. Where the indicator and protein have the same sign of charge, the protein error is usually small. If an indicator is adsorbed it tends to give the pH of the surface and not of the bulk. The difference may be as large as 2 pH units.

From scattered observations it appears that with indicators which are anions, the indicator error is large and positive on the acid side of the isoelectric point of the protein and small on the alkaline side. With indicators which are cations the relationship is reversed. This is the type of variation to be expected if the indicator error is mainly due to the adsorption of indicator at the protein surface. If the protein is positively charged it will tend to adsorb large organic anions, but not cations, and since the pH at the protein surface is greater than in bulk, the protein error is positive. Many other factors, such as the different valencies of the two forms of an indicator, and molecular interactions must also be involved.

The presence of proteins and lipids is a well recognized source of error in the colorimetric determination of pH. It was of interest to test the interference of protoplasmic and lipid substances of *Arbacia* eggs upon pH indicators. This was done in the following manner Wiercinski (1944). Eggs of the sea urchin *Arbacia* were strongly centrifuged and broken into hyaline and granular halves.

The respective halves were removed from the centrifuge tube with a pipette and put into separate test tubes. The portions were washed by shaking in 2 cc. of 0.35 M sodium citrate-citric acid buffer at the desired pH. This mixture was centrifuged at low speed and the supernatant fluid was decanted. Two cc. of fresh buffer was added. The proportion was approximately $^1/_{16}$ cc. of egg substance (hyaline or granular halves) to 2 cc. of buffer. The hyaline or granular buffer mixture was transferred to a small mortar and ground with a pestle. The ground mixture was tested electrometrically with a Beckman pH meter and colorimetrically with indicators. The colorimetric determination of the ground hyaline-buffer mixture was made on a white spot plate. The proportion was $^1/_{16}$ cc. of 0.02 per cent indicator to 1 cc. of mixture. The granular halves were too dark with pigment to permit proper estimates of their color.

The results are shown in Table I. Phenol red showed a colorimetric error of —0.1 to +0.02 pH units, bromcresol purple ranged from —0.17 to +0.22, and bromthymol blue +0.28 to +0.92.

Colorimetric pH error was also determined with dried *Arbacia* substance. This was added to a 0.1 N HCl solution diluted with distilled water to various pH intervals. The proportion was 5 cc. of acid and 0.05 grams of *Arbacia* substance.

The results are shown in Table II. Phenol red showed a colorimetric error of approximately +0.3 pH units, bromthymol blue approximately +0.1 to +0.4 and chlorphenol red approximately +0.2 to +0.4.

Table I. *Protein-lipid error of pH indicators determined with hyaline Arbacia substance* (WIERCINSKI 1944). Na citrate-citric acid buffer.

Indicator	Electrometric pH	Colorimetric pH	Electro-Color pH
PR	7.02	7.0	+ 0.02
PR	7.19	7.2	— 0.01
PR	7.30	7.3	0.00
BTB	6.48	6.2	+ 0.28
BTB	6.67	6.1	+ 0.57
BTB	6.83	6.1	+ 0.73
BTB	7.02	6.1	+ 0.92
BTB	7.19	6.5	+ 0.69
BTB	7.39	6.7	+ 0.69
BCP	5.64	5.5	+ 0.14
BCP	5.82	5.6	+ 0.22
BCP	6.04	6.0	+ 0.04
BCP	6.18	6.1	+ 0.08
BCP	6.48	6.1	+ 0.08
BCP	6.67	6.5	+ 0.17
BCP	6.83	7.0	— 0.17

Table II. *Protein-lipid error of pH indicators determined with dried Arbacia substance* (WIERCINSKI 1944). 0.1 N HCl solution.

Indicator	I Electrometric pH	II Colorimetric pH	I—II Electro-Color pH
PR	7.30	7.0	+ 0.3
BTB	6.43	6.3	+ 0.13
BTB	6.80	6.4	+ 0.4
CPR	6.12	5.7	+ 0.42
CPR	5.78	5.5	+ 0.28
CPR	5.72	5.5	+ 0.22

PR = phenol red, BTB = bromthymol blue, BCP = bromcresol purple, CPR = chlorphenol red.

d) Solutions containing very little electrolyte

KOLTHOFF (1937) has shown that with certain indicators (especially the sulfonphthaleins) there is considerable deviation from the true pH value in solutions that contain very little electrolyte. In solutions containing from 0.001 to 0.1 normal electrolyte the pH values have been found to be low by 0.20 to 0.45 units. This applies particularly to phenol red, chlorphenol red, bromphenol blue and bromcresol green.

e) Characteristics of an intracellular indicator

An acid-base indicator to be successfully used for cellular injection must not be toxic to the cell. It is also most important to inject the indicator into the clear hyaloplasm. This has been done by WIERCINSKI (1944). He injected indicators into the granule-free hyaline zone of centrifuged eggs. Following injection the dye rapidly diffuses through the cell. Small amounts of dye solution must be injected. If large amounts of solution in a droplet greater than $^1/_5$ the diameter of the cell are injected, then a membrane is seen to form about the injected droplet. If the injection is continued the cell is disorganized due to the continued pressure, and cytolysis will result.

Indicator solutions may be immiscible with protoplasm and membranes can then be seen to form about the droplets of indicator solution. When cytolysis occurs the protoplasm becomes more acid. The method of microinjection, when properly used, appears to be decidedly better than the immersion of cells in vital dyes.

f) Mixed indicators

A mixed indicator has the advantage of a sharp color transformation point (p^T) at a given pH, KOLTHOFF (1937). Such indicators were used by WIERCINSKI (1944) for microinjection into the hyaline zone of *Arbacia punctulata* eggs. The color characteristics of the mixed indicator standards are as follows:

Bromcresol purple — bromthymol blue, pT at pH 6.7. The useful range is from pH 5.7 to 7.3 at 0.2 pH intervals. At pH 7.3 the color reaction is greenish blue. The blue color diminishes at pH 7.1 and 6.9. From pH 6.7 to 6.1 the color is predominately green. At pH 5.9, the color is yellowish green and a deeper green at pH 5.7.

Bromcresol green — chlorphenol red, pT at pH 6.1. The useful range is from pH 6.9 to 5.3 at 0.2 pH intervals. Above pH 6.1 the color increases in violet intensity. Below pH 6.1 the color appears green and at pH 5.3 a yellowish green.

Chlorphenol red — aniline blue pT at pH 5.8. The useful range is pH 6.6 to 5.0 at 0.2 pH intervals. Above 5.8 the color is increasingly blue. At pH 5.8 the reaction is blue green and below pH 5.8 the color is increasingly green.

Bromcresol green — sodium alizarine sulfonate pT at pH 5.6. The useful range is from pH 6.4 to 4.8 at 0.2 pH intervals. On the alkaline side of the transformation point, this indicator exhibits a violet hue at pH 5.6 and is increasingly blue to pH 6.4. On the acid side, the color appears to change from a green to a yellow green at pH 4.8.

g) Measurement of micro-quantities

The concentration and quantity of the indicator must be known. It is possible to determine the exact amount of fluid that is injected by filling the pipette with a known quantity of fluid from a capillary tube the volume of whose content can be determined before and after the pipette has been filled. This is the technique suggested by BENEDETTI-PICHLER (1937, 1940).

CORSON (1943) controlled the rate of injection (using a Chambers micromanipulator) and measured the volume of the solution between 2 droplets of oil of different specific gravities. The droplet of light oil at the microtip was injected together with the solution and served as an indicator of the actual injection of the solution into the cell. It also prevented the mixture of the solution in the micropipette with that of the hanging drop in which the living cells were suspended.

The quantity of fluid to be injected was measured by placing the pipette in a plane perpendicular to that of the long axis of the microscope which was fitted with a calibrated ocular micropipette. All of the solution thus drawn into the pipette was injected into the cell. The micropipettes were calibrated by immersion in cedar oil. The results were checked by injecting aqueous solutions into an oil of the same specific gravity and then measuring the diameters of the droplets.

h) Colorimetric measurement for tissue fluids

HOMER (1917) working with the colorimetric determination of hydrogen ion concentration in many samples of sera found that the large personal error involved indicates that it is advisable to check the values obtained against the electrometric method. For example, in plasma containing

6 to 8% protein the sulfonphthalein indicators will give an error of 1 to 2 units of pH when the colorimetric method is compared with the electrometric measurement.

Brown (1923) developed a colorimetric method for measuring the pH of small quantities of fluid accurate to within 0.1 pH. Glass cells were made by cementing cover slips to one edge of small glass rings with finely ground edges. These were matched with buffers and indicators at known pH's.

i) Crushing cells in indicator solutions

Vlès (1924) and Reiss (1924 a, b) crushed cells in order to determine the pH of their contents. The method of introducing indicators by means of crushing presents problems related to the injury of the cell. The first stage is to burst the cell by pressure on the coverslip which is over the cell. Then by releasing the pressure upon the coverslip the intake of indicator into the cell occurs. The following figures illustrate this procedure. Fig. 3 and 4.

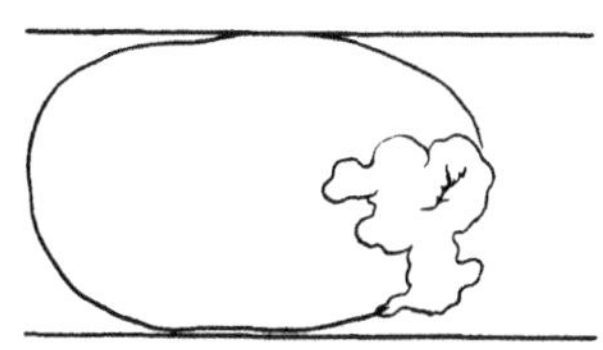

Fig. 3. Method of crushing. First stage: Bursting of the cell by compression. Redrawn from Reiss (1926 a).

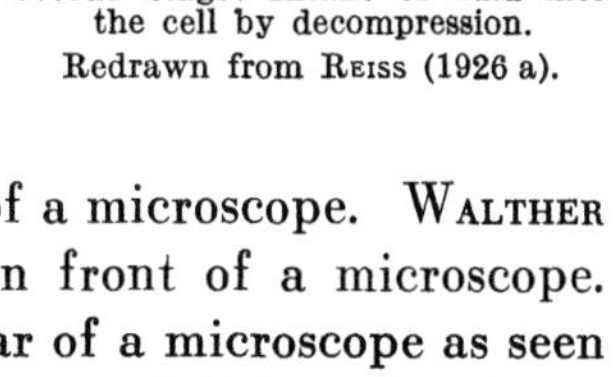

Fig. 4. Method of crushing. Second stage: Intake of fluid into the cell by decompression. Redrawn from Reiss (1926 a).

Vlès, Achard and Prikelmaier (1923) crushed *Paracentrotus* eggs at — 15° C. and reported a pH of 5.8 for the cytoplasm. At low temperature the CO_2 content of the protoplasm would be retained to a larger degree than at a higher temperature. Tchakotine (1929) confirmed Vlès' value for the pH of sea urchin egg protoplasm.

Reiss (1926 a, b) used a colorimeter as shown in Fig. 5. This apparatus is also known as the Bjerrum wedge, (Bjerrum 1917). Gillespie (1920) used a comparator block for test tube solutions. Pantin (1923 a, b) and Needham and Needham (1925 a, b, 1926) set a series of test tubes in front of a microscope. Walther and Ulrich (1926) used a series of microtubes in front of a microscope. Vlès (1926) used a microcolorimeter over the ocular of a microscope as seen in Fig. 6. Vlès and Gex (1928) and Vlès and Vellinger (1928) used a direct vision prism spectroscope in conjunction with a microscope for studying the pigment within living cells.

j) Microinjection of indicator solutions into living cells

The method as originated by Kite (1913) and perfected by Chambers and collaborators (1922) has been widely used in the determination of intracellular pH.

The principle of microinjection is illustrated in Fig. 7.

Kite (1913) microinjected indicators into the interior of *Amoeba proteus* to determine a possible relation between an excess of H^+ and OH^- ions.

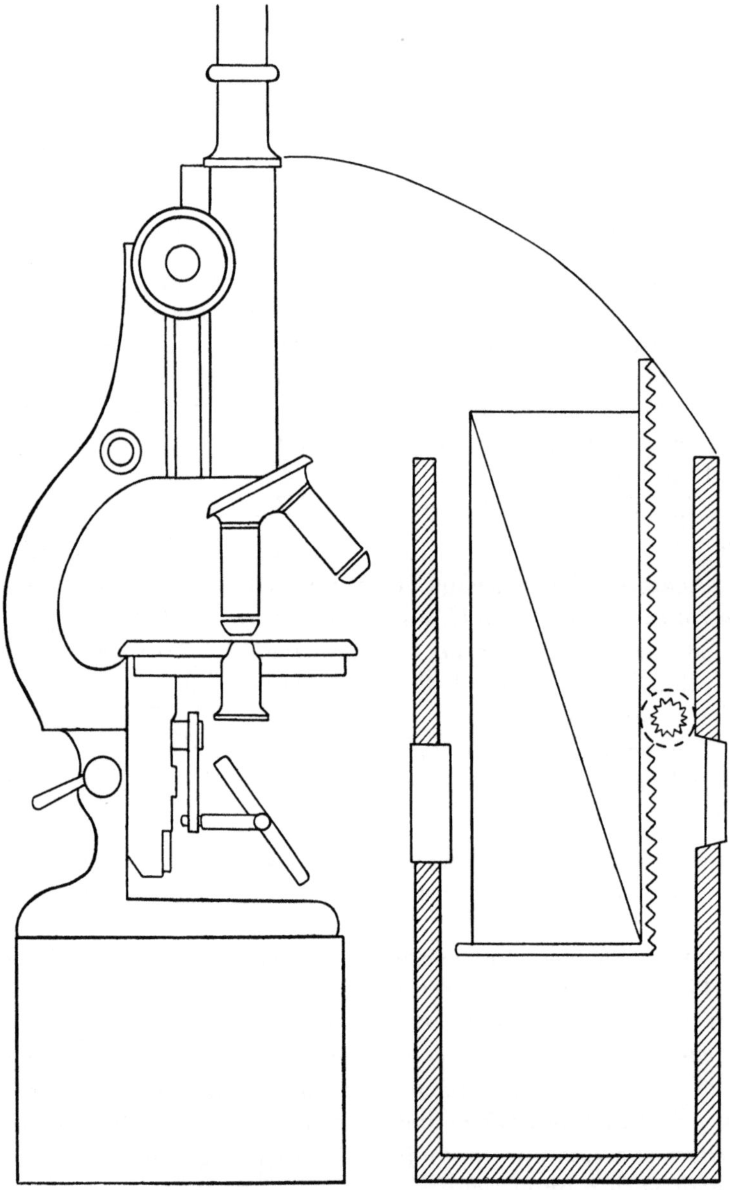

Fig. 5. Colorimetric apparatus for the determination of intracellular pH. Redrawn from Reiss (1926 a, b).

Azolitmin, sodium alizarin sulfonate, tropeolin 000 No. 1, methyl orange and Congo red dissolved in from ½ to ¾ molar cane sugar were used. A neutral to slightly alkaline reaction was shown by all the indicators.

Needham and Needham (1925 a, b) and (1926) used the method of Pantin (1923 a, b) to observe the tint of an indicator in the field of a compound microscope. A row of test tubes filled with buffer solutions containing indicator was interposed between the light source and the microscope. The images of the tubes were thrown on to the microscopic field and their

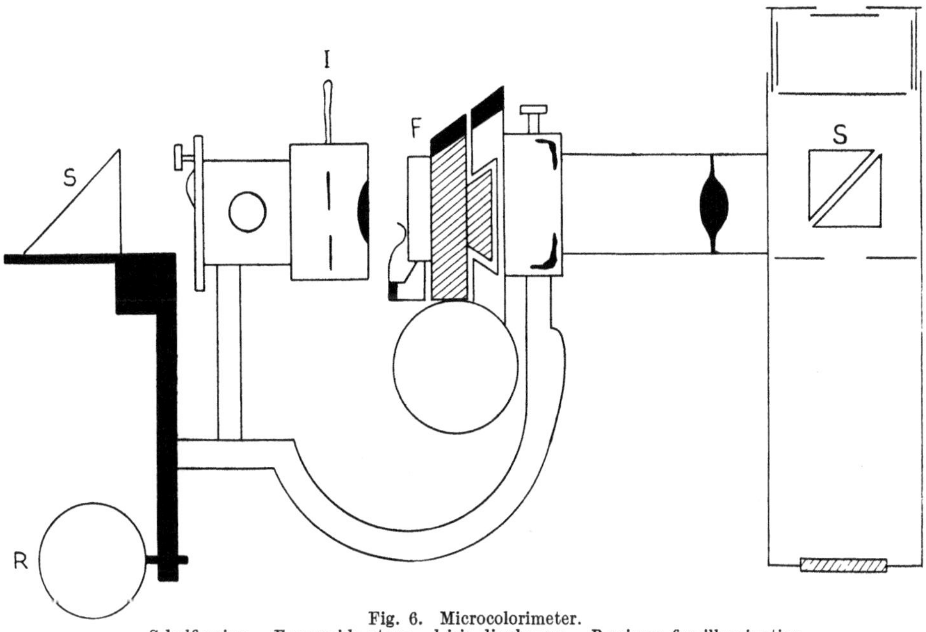

Fig. 6. Microcolorimeter.
S half prism. *F* movable stage. *I* iris diaphragm. *R* mirror for illumination.
Redrawn from VLÈS (1926).

tints were compared with that in the interior of the amoeba. These have a negligible protein error according to HOMER (1917) and CLARK and LUBS (1928).

KOPAC (1935) showed that when a probing object is inserted into an animal cell a layer of protoplasm separates the object from the rest of the cell.

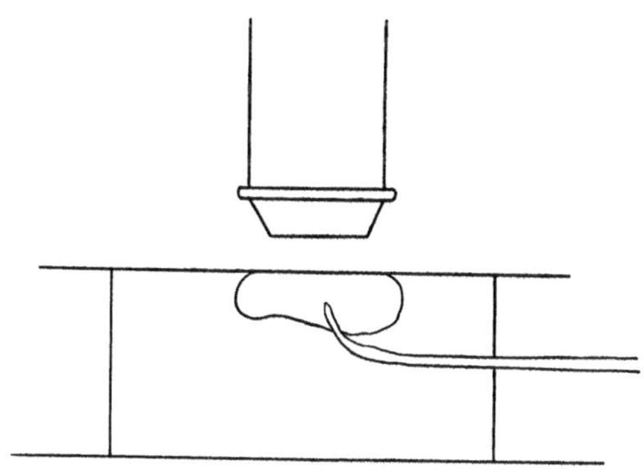

Fig. 7. Method of microinjection.
Redrawn from REISS (1926 a).

Other investigators, RAPKINE and WURMSER (1926, 1928), CHAMBERS (1949) and collaborators have injected pH indicators into cells by means of micropipettes. The color changes are clearer than with vital dyes, since all of the cell becomes tinted, and the range of possible dyes is very much extended, but these indicators are generally toxic for the cell and the introduction of a pipette produces an injury which may affect the pH.

k) Microinjection of indicator particles into living cells

SCHMIDTMANN (1924, 1927 b), SCHMIDTMANN and MATTHES (1927) and SCHMIDTMANN and SEKI-NUREDDIN (1927) injected minute granules of indicator dyes into the protoplasm of mammalian tissues. They obtained small fragments of tissue from teased preparations and from frozen sections. The color produced by the dissolved granule was compared with that resulting from a similar granule placed in a droplet of known buffer solution.

OGAWA (1928, 1929) reexamined the SCHMIDTMANN method, GRAFF (1924), of determining the hydrogen ion concentration by inserting into the cell cytoplasm particles of coloring matter. The cell must be as fresh as possible. When it is partially broken, the cell is rendered acidic and unfit for examination. OGAWA also made a table of color tones, the use of which is claimed to afford a simpler and more exact pH determination.

l) Indicators injected into the blood stream of mammals

The work of ROUS and associates (1924—1927) is of particular importance. His observations indicated that acid-base conditions between blood and tissue, in general, are such that the tissues are less alkaline than the blood. In some organs a notable acidity prevails. ROUS' method was to inject the indicator into the blood stream. After a suitable interval the animal was sacrificed under oil.

ROUS found that the sodium salts of the sulfonphthalein indicators stain living tissues better than litmus. Several are well tolerated as the survival of the animal after injection with the indicator has shown. These are thymol blue, cresol red, and phenol red. Bromthymol blue stains tissues poorly and produces an intoxication in mice. Bromcresol purple stained tissues intensely but the dose necessary was lethal. Methyl red was easily reduced in the living animal.

Bromcresol green possesses much the same range as methyl red. However, it was too far to the acid side to be useful. ROUS (1925 a) found that litmus was stored in the granules of the reticulo-endothelial elements. There was a notable acidity within the cell granules. ROUS (1925 b) found that macrophage granules stained, in vivo, with sulfonphthalein indicators showed a pH of 3.0 or less. ROUS (1925 c) also found that on exposure to air, without disturbance of circulation, some of the tissues become more alkaline due to the escape of carbon dioxide. According to ROUS (1925 d), "acid" organs are those that have a greater metabolic activity. All tissues examined were relatively acid compared to the blood plasma. Errors of interpretation may arise from the fact that the phthaleins yield unreliable results in the presence of proteins.

ROUS (1926) studied ischemic mouse tissue, staining it vitally with various indicators. When tissues are very acid they can be implanted and give vigorous grafts. Injured tissues tend to to be as alkaline as the surrounding blood; these can not readily be implanted.

ROUS and BEATTIE (1926) studied rat tissues. They found that if hydrochloric acid and sodium carbonate are injected into the blood stream there is a marked alteration in the color of certain of the matrix tissues stainable, in vivo, with sulfonphthalein indicators. Lactic acid was well tolerated. Salt solution and sugar in large amounts cause no change in color such as would indicate changes in reaction.

The best that one can do is to use those indicators that are least affected by proteins and salts. DRURY and ROUS (1926 a) perfused organs with water, salt, or lymph of the animal and extracts of the organs were obtained. The turbid extracts were colored with phthalein, the reactions determined colorimetrically with the aid of an arc light. The findings were compared with potentiometer readings upon the same material. The results of these two methods were in close agreement. No evidence was obtained of indicator errors in the association of phthalein with tissue substances, DRURY and ROUS (1926 b). The colors seen in vitally stained tissues are indicative of a notable acidity. There is no indication of the influence of tissue activity upon the coloration.

MARGARIA (1934) found an apparent change of pH on stretching a muscle. An isolated frog sartorius muscle, colored with an indicator (bromcresol purple, or bromthymol blue) previously injected into the animal, showed during contraction, a change of color, which, if it were due to an alteration of pH, corresponded to a change of about + 0.5. This change occurred in twitch or tetanus, isometric or isotonic, and was observed by the eye, it set in with contraction and disappeared with relaxation. Also, see ARVANITAKI and CHALAZONITIS (1951).

m) Intraperitoneal injection

HATANO and IWATA (1933) introduced pH indicator dyes intraperitoneally into full grown mice and white rats to stain the histiocytes. Phthalein dyes were used including the sulfonphthaleins. The pH within the cytoplasm was found to vary from 5.2 to 6.8. In vitro cultivation of the histiocytes showed that the cytoplasmic reactions turned alkaline.

n) Absorption of indicator by lower organisms

CROZIER (1923) used bromthymol blue on *Paramecium* and *Opalina.* He observed pH values of 6.7 for the protoplasm of *Paramecium* and 6.2 for *Opalina.* The reaction of the indicators to alkalies (NaOH and NH_4OH) shows that the indicator is not merely condensed on the external surface of the organism. He concluded that protoplasm is, general, significantly acid in character, rather than alkaline or neutral.

NOMURA and KOKUBO (1934) working with the hydromedusa, *Aglantha digitalis,* found that phenol red and bromthymol blue stained the tissues of this organism. They allowed the organism to swim in a solution of these sulfonphthalein indicators. Bromthymol blue also gave satisfactory results with the ciliated epithelium of *Pecten yessoensis.* They observed the pH of the tissue at 7.0.

4. Natural indicators in living cells

a) Observations with the microscope

The work of CROZIER (1918) with *Porifera* is important in that he observed natural indicators found in animal tissues. The color changes are sufficient to afford some indication of the normal intracellular acidity. The method used was to determine acidities at the points of color transitions by adding a small volume of the concentrated pigment extract to buffer mixtures of known acidity. Another way was to determine the hydrogen ion concentration in which the respective pigments appeared to be neutral. These observations were made with indicators.

b) Spectrophotometric method

VLÈS and VELLINGER (1928) compared the spectrum of the pigment within the living egg of *Arbacia aequituberculata* with that of alcoholic extracts of the pigment in contact with various buffer solutions. He concluded that there exists in the egg "points" which have the pH value of 5.5 with a variability of ± 0.3. Actually, VLÈS and VELLINGER observed the pH of the pigment granules and not of the cytoplasm. It is interesting to note that this value agrees with that of VLÈS (1924) and REISS (1924 a) obtained by crushing eggs in indicator solutions. This method of VLÈS and VELLINGER is based on the spectrophotometric determination of the absorption of two bands characteristic for the two forms of the indicator. Thus, it is possible to obtain the pH independently of the concentration of the indicator.

C. Methods of Calculation

1. Henderson-Hasselbalch equation

FENN (1928) and FENN and MAURER (1935) by means of data calculated with the Henderson-Hasselbalch equation, claim that the pH of frog muscle protoplasm is 6.9. CONWAY and FEARON (1944) studied acid-labile carbon dioxide in mammalian skeletal muscle. The estimations made on the basis of these determinations indicate that the pH within the resting muscle fiber is around 6.0. They state that the hydrogen ion concentration within the muscle cell depends, to an extent related to the buffering power of the muscle protoplasm on the intracellular fluid, the composition of which in turn reflects the adequacy of the blood supply to the muscle. Also, see STELLA (1929) and COWAN (1933).

2. The Zeta potential

ALBERT (1952) suggests the electro-kinetic, or Zeta potential for the measurement of pH on biological surfaces. HARTLEY and ROE (1940) derived the following equation.

$$\text{pH (surface)} = \text{pH (solution)} + \frac{\zeta}{0.06}$$

(where the Zeta potential is measured in volts).

The great majority of dyes tend to congregate at surfaces; hence the limitation of indicators when used to determine the pH inside of cells, ALBERT (1952).

3. Buffering Power

BATE-SMITH (1948) calculated what he terms "basal pH" from the lactic acid and pH values for the muscle of animals such as rat, pig, steer, and lamb. His formula is

$$\frac{\delta B}{\delta \text{pH}} = \text{Buffering power of muscle.}$$

δB = expressed as milliequivalents of lactic acid per 100 grams of muscle.

V. Tabulated Results for the Animal Kingdom

The results for data on the pH of animal cells is presented in phylogenetic sequence. The existing data is given for each phylum, with a tabular notation for the author(s) and the method(s) that they used. In order to save space, symbols are used to denote parts of cells and tissues. The symbols are as follows:

Symbols used in the following tables

(AH)	animal hemisphere	(IS)	intercellular staining
(B)	blood	(L)	liver
(BF)	body fluids	(M)	macronucleus
(C)	cytoplasm	(MF)	muscle fiber
(DE)	Deep immersion of electrodes	(MS)	striated muscle
(En)	endoplasm	(NI)	natural indicators
(E)	erythrocyte	(NR)	neutral reaction
(Ex)	exoplasm	(N)	nucleus
(FV)	food vacuole	(S)	serum
(G)	granules	(SA)	slightly acid
(H)	hyaline cytoplasm	(SL)	slightly alkaline
(HC)	hyaline cytoplasm made granule-free by centrifugation	(Ps)	pseudopodia
		(T)	tissue
(HP)	hepatopancreas	(V)	vacuoles
(I)	injured cytoplasm	(VH)	vegetative hemisphere

		pH	Method	Author(s)
A. Invertebrata				
1. PROTOZOA				
a) Sarcodina				
Amoeba dubia	C	6.9 ± 0.1	Microinjection of Clark indicator solution	CHAMBERS, POLLACK, and HILLER (1927)
	C	6.8		CHAMBERS (1929 a, b)
	C	7.6		NEEDHAM and NEEDHAM (1925 a, b)
	C	NR-SL		KITE (1913)
	C	7.3		
	C	6.8 ± 0.2		WIERCINSKI (1944)
	C	6.9		COHEN, CHAMBERS, and REZNIKOFF (1928)
	C	6.0		SPEK and CHAMBERS (1933)
	C	5.0		
	H	6.8 ± 0.2		WIERCINSKI (1944)
	N	7.5 ± 0.1		CHAMBERS, POLLACK, and HILLER (1927)
	I	5.3 ± 0.2		
	I	5.0 ± 5.3	Crushing cells in indicator solution	REISS (1924 a, b, 1928)
	V	7.6	Vital staining with basic dyes	SPEK and CHAMBERS (1933)
Amoeba proteus	C	6.8 ± 0.2	Microinjection of Clark indicators	WIERCINSKI (1944)
	C	7.6		NEEDHAM and NEEDHAM (1925 a, b)
Resting	C	6.7—7.3		
	N	6.7—7.6	Vital staining with basic dyes	DATE (1931)
After movement	C	6.5—7.0		
	C	6.8—7.0	Microinjection of Clark indicator solution	CHAMBERS, POLLACK, and HILLER (1927)
	H	6.8 ± 0.2		WIERCINSKI (1944)
	I	5.0		NEEDHAM and NEEDHAM (1925 a, b)
	N	6.7—7.6		CHAMBERS, POLLACK, and HILLER (1927)
Amoeba limax	E	7.6—7.8	Vital staining with neutral red.	PANTIN (1923 a)
	Ex	7.2		
Active	Ps	6.8		
Resting	Ps	7.0—7.2		

		pH	Method	Author(s)
Amoeba polypodia	H	7.2—7.3	Microinjection of Clark indicator solution	SPEK and CHAMBERS (1933)
Entamoeba histolytica:				
Human	C	7.0	Indicators added to ten day old culture	YOSEZATO (1933)
	I	SL		
Kitten	C	7.0		
	I	SL		
Resting	En	6.5	Microinjection of Clark indicator particles	OGAWA (1929)
After movement	En	5.8—6.3		
Entamoeba coli:				
Resting	En	6.5		
	Ex	6.8		
After movement	En	5.8—6.3		
	Ex	6.0—6.5		
	N	7.0		
Gromia sp.	C	5.3	Crushing cells in indicator solutions	VLÈS (1924), REISS (1926 a)
b) Infusoria				
Paramecium caudatum:				
In aerobiosis	C	6.7—6.9	Vital staining with basic dyes	GERSCH (1936, 1937)
In anaerobiosis	C	8.0		
Digestive vacuole		4.0	Vital staining with acid pH indicators	SHAPIRO (1927)
	N	6.7	Natural indicators	CROZIER (1923)
Paramecium:				
Food vacuole	FV	3.0	Ingested yeast cells stained with Congo Red	BOZLER (1924)
	FV	1.0		NERENSTEIN (1925)
	FV	cyclic: 7.6 to 4.0 to 7.0	Litmus, Congo Red and Phenol red	SHAPIRO (1927)
	FV	5.6 > 4.0—8.0	Neutral red	MAST (1947)
		3.2	Congo Red	
		1.4	Thymol blue	
Cryptochylum (Cryptochilidium) nigricans	C	5.0	Crushing cells in indicator solution	REISS (1924, 1926 a)

		pH	Method	Author(s)
Opalina ranarum:	N	6.7	Natural indicators	Grozier (1923)
	C	7.6	Clark indicator solution	Kedrowsky (1931)
Natural medium	C	6.0—6.2	Unboiled, crushed and ground cell suspension	
Phosphate buffer:				
pH 9.0—8.1		5.9—6.0		
pH 8.1—6.2		6.2—6.4		
pH 6.2—5.2		6.1—6.4		
Vorticella sp.				
At resting	V	5.0	Vital staining with basic dyes	Pastori (1932)
After contraction	V	3.5		
Digestive	V	4.5	Vital staining with acid pH indicators	Shapiro (1927)
Nyctotherus cordiformis:				
In aerobiosis	C	7.1	Microinjection of Clark indicator solution	Needham and Needham (1925 b)
In anaerobiosis	C	7.1		
	C	6.8—7.0	Vital staining with basic dyes, 0.025%	Nassanov (1932)
	M	6.0	Neutral red	
Inclusion bodies				
Pulsating Vacuole		7.0—7.4		
Data at following same as for *Nyctotherus:*				Matsuda (1936 a, b)
Balantidium duodeni				
Ostracodium triloricatum				
Endipodium magii				
Polyplastron magii				
Entodinium caudatum				
Entodinium longinucleatum				
c) Mastigophora				
Trichomonas vaginalis	C	5.5—6.8	Vital staining with acid pH indicators and basic dyes	Matsuda (1936 a, b)
Leishmania donovani	C	6.4—6.8	Vital staining with acid pH indicators and basic dyes	Saito (1936)

		pH	Method	Author(s)
d) Sporozoa				
Actinosphaerium eichorni				
Digestive vacuole:				
Before taking food		6.6—6.9	Microinjection of Clark indicator solution	
During digestion		4.3		HOWLAND (1928, 1930)
After digestion		5.4		
2. PORIFERA				
Spongida				
Aplysinidae				
Body substance	NI	6.0	Natural indicators	CROZIER (1918)
Thin projections	NI	8.0		
Cortex	NI	8.0		
Choanosome	NI	7.0		
Mellita sexies-perforata				
Integument	NI	7.5		
3. COELENTERATA				
Sertularia pumila	T	5.2	Crushing cells in indicator solutions	REISS (1924 a, b)
Hydractinia echinata	T	5.2		
Sagartia parasitica	T	5.6—6.0		
Hydrozoa				
Hydra fusca				
Ectoderm:				
Cortical protoplasm		7.5—7.6	Microinjection of Clark indicator particles	SCHMIDTMANN (1924, 1925 a, b)
Inner protoplasm		6.7—7.0		
Nucleus	N	6.8		
Nematocyst cell	C	7.0—7.2		
Cnide		7.8		
Pigment cells	C	7.4—7.6		
Endoderm:				
Digestive cells,				
after 3 day fast	C	7.2		
after injection	C	6.1		
Clytia johnstoni			Vital staining with acid pH indicators	ATKINS (1922)
Medusa stage	T	6.6		
Tiara pileata:				
Medusa stage	T	6.6		
Mantle, tentacle, canal		6.2—6.5		
Umbrella		7.2		

		pH	Method	Author(s)
Aglantha digitalis			Vital staining with acid pH indicators sulfonphthaleins	NOMURA and KOKUBO (1934)
Medusa stage				
Umbrella epithelium and mesoglea	T	7.0		
4. ECHINODERMATA				
Paracentrotus lividus:				
unfertilized egg	C	7.2 ± 0.2	Microinjection of Clark indicator solution	RAPKINE and WURMSER (1926)
fertilized egg	C	6.6 ± 0.1		
Strongylocentrotus lividus (also Para):				NEEDHAM and NEEDHAM (1925 a, b, 1926)
unfertilized egg	C	6.5	Crushing cells in indicator solution	REISS (1924 a, b)
	C	5.5		VLÈS (1924)
unfertilized egg		5.1	Electrometric measurement of crushed cells and tissues	VLÈS, REISS, and VELLINGER (1924 a, b, c)
unfertilized egg		5.8—5.9		VELLINGER (1926)
unfertilized egg	C	6.6 ± 0.1	Microinjection of Clark indicator solutions	NEEDHAM and NEEDHAM (1925 a, b, 1926)
fertilized egg	C	5.4—5.6	Crushing cells in indicator solutions	REISS (1924 a, 1925, 1926 a, b, 1928 b)
unfertilized egg	C	7.0	Microinjection of Clark indicator solutions	RAPKINE and WURMSER (1926)
	N	7.0		
unfertilized egg	C	5.8		TCHAKHOTINE (1929)
Echinicardium cordatum:				
Unfertilized egg	C	5.0	Crushing cells in indicator solution	REISS (1924, 1926)
unfertilized egg	C	6.6	Microinjection of Clark indicators	NEEDHAM and NEEDHAM (1925 a, b, 1926)
unfertilized egg	C	6.6 ± 0.1		
Arbacia punctulata:				
Unfertilized egg				
Pigment		5.3 ± 0.3	Spectrophotometric pH determination of natural indicator	VLÈS and VELLINGER (1928)
Normal	C	6.8 ± 0.2	Microinjection of Clark indicator	PANDIT and CHAMBERS (1932)
Cytolysis	C	5.3 ± 0.2		

		pH	Method	Author(s)
Pigment		5.5		
	N	7.6—7.8		
Arbacia punctulata:				
unfertilized egg	HC	6.2 ± 0.2	Microinjection of Clark indicators	WIERCINSKI (1941, 1944)
Arbacia aequituberculata	C	5.0—5.2	Electrometric measurement of living cells with the hydrogen electrode	VELLINGER (1926)
Asterias glacialis:				
unfertilized egg	C	6.6 ± 0.1	Microinjection of Clark indicator solution	NEEDHAM and NEEDHAM (1925 a, b, 1926)
Oocyte	C	7.2 ± 0.2		RAPKINE and WURMSER (1926)
Asterias rubens:				
unfertilized egg	C	7.2 ± 0.2		
	N	7.0		
Asterias forbesii	C	6.7 ± 0.1		CHAMBERS (1928)
	N	7.5 ± 0.1		CHAMBERS and POLLACK (1920) CHAMBERS, POLLACK, and COHEN (1929)
fertilized egg				
4 blastomere stage	C	6.7 ± 0.1		
	N	7.5 ± 0.1		
cortical protoplasm	C	5.0	Vital staining with basic dyes	SPEK 1933 SPEK and CHAMBERS (1933) SPEK (1934 b, c)
inner protoplasm	C	7.6		
Ophiura lacertosa:				
unfertilized egg	C	6.75 ± 0.1	Microinjection of Clark indicator solution	NEEDHAM and NEEDHAM (1925 a, b, 1926)
Psammechinus miliaris	C	5.2	Crushing cells in indicator solution	REISS (1926 a)
5. MOLLUSCA				
Trochocochlea lineata	C	5.2	Crushing cells in indicator solution	REISS (1924 a—1928 a)
T. hepatopancreas	T	5.6—6.0		
Aplysia limacina	C	4.7		
	C	5.0—5.6		
Chromodoris zebra	NI	5.6—7.6	Natural indicators	CROZIER (1918)
Octopus vulgaris	L	7.0	Electrometric measurement of crushed cells and tissues	DUVAL, GUEYLARD and PORTIER (1925)
	B	7.7		

		pH	Method	Author(s)
Sepia officinalis	L	6.0		
	B	7.7		
Anodonta cygnea	L	6.8		
	B	7.8		
Mactra solidissima	C	6.8 ± 0.2	Microinjection of Clark indicator solution	GRAND (1938)
	N	7.4 ± 0.2		
	I	5.4 ± 0.2		
Patella picta: muscle	T	5.6—6.0		DUVAL, GUEYLARD, and PORTIER (1925)
Pecten yessoensis: ciliated cells of gills	C	7.0	Vital staining with basic dyes	NOMURA (1939)
6. ANNELIDA				
Chaetopoda				
Sabellaria alveolata	C	5.0	Crushing cells in indicator solution	REISS (1924 a, b, 1928 a)
	N	7.2—7.5		
After maturation	C	7.8—12.0	Vital staining with basic dyes	REISS (1926 a), FAURÉ-FRÉMIET (1923)
Unfertilized egg	C	6.6 + 0.1	Microinjection of Clark indicator solution	NEEDHAM and NEEDHAM (1925 a, b, 1926)
Spirographis sp.	C	6.0	Crushing cells in indicator solution	REISS (1924 a, b, 1928 a)
Arenicola claparedii	N	7.2—7.5	Vital staining with basic dyes	SPEK and CHAMBERS (1933)
Nereis dumerlii	V, G	5.0 to 8.0	Metachromatic error	SPEK (1934 a, b, and c)
Nereis limbata	V, G	5.6 to 8.0	Metachromatic error	SPEK (1934 a, b, and c)
	C	5.6—6.0	Crushing cells in indicator solution	REISS (1926 a)
Lumbricus		7.3	Ectoderm indicator particles	SCHMIDTMANN (1925 b)
7. HIRUDINEA				
Platydella soleae			Crushing cells in indicator solution	REISS (1926 a)
salivary gland	T	5.6—6.0		
clitellum	T	5.6—6.0		
8. ARTHROPODA				
a) Crustacea				
Chondracanthus (egg)	C	5.0	Crushing cells in indicator solution	REISS (1926 a)
Maja (maia)	C	5.0		
squinado (egg): muscle		7.1	Carbon dioxide dissociation curve	COWAN (1933)

		pH	Method	Author(s)
Astacus fluviatilis	HP	6.0—6.3	Electrometric (1) measurement of 3 gms. of mashed liver in 5 cc. of distilled water	DUVAL, GUEYLARD, and PORTIER (1925)
	B	7.7	Electrometric (2) measurement of living cells with the hydrogen electrode	
Platycarcinus pagurus	HP	5.9—6.8	Electrometric measurement of 3 gms. of mashed liver in 5 cc. of distilled water	
	B	7.7	Electrometric measurement of living cells with the hydrogen electrode	
Portunas puber	HP	7.2	Method marked (1) as above	
	B	7.7	Method marked (2) as above	
Carcinus maenus	HP	6.1	Method marked (1) as above	
	B	7.7	Method marked (2) as above	
Carcinus maenus	MF MS	6.9	Micro-glass electrodes	CALDWELL (1954)
Cancer, Carcinus, Homarus, Eupagurus	MS	7.05	Electrometric measurement of tissue frozen in liquid air	HOET and CARRIDGE (1926)
b) Insecta				
Dytiscus marginalis	MS	6.7—6.8	Microinjection of Clark indicator particles	SCHMIDTMANN (1925 b)
Chironomus plumosis larva salivary gland cell	C	7.1—7.2	Vital staining with basic dyes	PARAT (1928)
	C	7.2 ± 0.2	Microinjection of Clark indicator solution	RAPKINE and WURMSER (1927)
	V	6.8—6.9	Vital staining with basic dyes	
	I	5.0		ALEXANDROV (1932)

		pH	Method	Author(s)
Chironomus				
in aerobiosis	C	7.2		
in anaerobiosis	C	6.2		
	N	7.0—7.2		
	C	7.0	Microinjection of Clark indicator solution	BERTANI (1944)
	N	5.8—6.2		
Calliphora erythrocephala (larva)				
salivary gland cell	C	7.2 ± 0.2		RAPKINE and WURMSER (1927)
Leptinotarsa decemlineata egg, larva and pupa	BF	5.9—6.8	Colorimetric method on mashed tissues and body fluids	FINK (1925)
Cotinis nitida egg and larva	BF	6.2—7.3	Colorimetric method on mashed tissues and body fluids	FINK (1925)
Popillia japonica, egg	BF	7.1		
Epilachna borealis				
egg, larva, pupa	BF	5.9		
Ancylis comptana				
larva and pupa	BF	5.9—6.8		
Hylemia cilicrura				
egg, larva, pupa	BF	5.9—7.1		
Anasa tristis				
egg, larva, pupa	BF	6.2—6.8		
Pieris rapae				
larva and pupa	BF	5.9—6.4		

Body Fluids FINK (1925), GLASER (1925)

Species	Egg		Larva		Pupa	
	Early pH	Late pH	Early pH	Late pH	Early pH	Late pH
Leptinotarsa decemlineata	6.8	6.8	6.8	5.9	5.9	6.8
Cotinis nitida	6.2	7.1	7.3	—	—	—
Popillia japonica	7.1	7.1	—	—	—	—
Ancylis comptana	—	—	6.8	5.9	5.9	6.8
Epilachna borealis	5.9	5.9	5.9	5.9	5.9	5.9
Hylemia cilicrura	5.9	6.6	7.1	6.6	5.9	6.6
Anasa tristis	6.2	6.4	6.6	6.6	6.8	6.8
Pieris rapae	—	—	6.4	6.2	5.9	6.4

		pH	Method	Author(s)
Bombyx mori				
Egg cells in ovary	I	6.58	Quinhydrone and calomel electrode measurement of crushed cells and tissues	Ongaro (1931 a, b)
Oviposited eggs				
unfertilized	I	6.58		
fertilized	I	6.50		
hibernal	I	6.5—6.7		
incubated	I	7.0		
c) Diptera				
Larval blood				
Drosophila melanogaster	B	7.10	Capillary glass electrode	Boche and Buck (1942)
Sciara coprophila	B	7.12		
Chironomus tentans	B	7.48		
Drosophila melanogaster	B	7.12		
Sciara coprophila	B	7.38		
Buffer solution		5.662 Av.		
Melanoplus differentialis				
Grasshopper				
male nymph	B	7.126 Av.	Microelectrode and vessel	
male adult	B	7.20 Av.		
female adult	B	7.25 Av.		
Encoptolophus sordidus				
male adult	B	6.95 Av.	Microelectrode and vessel	
Popilla japonica				
Japanese beetle larva				
3rd instar	B	7.30		
Leptinotarsa decemlineata (Potatoe beetle)				
male adult	B	6.0		
B. Chordata				
1. UROCHORDATA				
Tunicata				
Ascidia mentula	C	5.0	Crushing cells in indicator solution	Reiss (1926)
	N	7.2—7.5		

		pH	Method	Author(s)
Ascidia mentula				
Unfertilized	C	6.6 ± 0.1		
	C	6.8	Microinjection of indicator solution	NEEDHAM and NEEDHAM (1925, 1926)
	V	5.0	Vital staining with basic dye	PARAT (1928)
Cerebral ganglion	T	5.0	Crushing cells in indicator solution	PARAT (1928)
Styelopsis sp.	C	4.7		REISS (1926 a)
Fragarum elegans	T	5.6—6.0		
Sabellaria alveolata	C	6.6	Microinjection of indicator solution	NEEDHAM and NEEDHAM (1925, 1926)
2. VERTEBRATA				
a) Osteichthyes				
Teleostei				
Fundulus heteroclitus	C	6.1—6.8 6.39	Quinhydrone and calomel electrode measurement	ONGARO (1931)
	I	4.4	Microinjection of Clark indicator solution	ARMSTRONG (1927)
	C	6.6—6.8		CHAMBERS (1932)
Cell in cytolysis	I	5.2—5.6		CHAMBERS (1926)
	MS	6.6—6.8	Electrometric measurement of crushed cells and tissues	BENSON (1928)
Fundulus:				
4 to 6 days after fertilization			Microinjection of Clark indicator solution	ARMSTRONG (1927)
Sub-chorionic space:				
in sea water		8.2		
Pericardial space		8.2		
In distilled water 18 hours:				
Sub-chorionic space		5.6 ± 1.0		
Pericardial space		8.2		
Fundulus:				
Yolk, average value		6.39	Hydrogen microelectrode	BODINE (1927)
Oryzias latipes:				
1 to 4 cell stage, yolk		6.4 ± 0.2	Microinjection of sulfonphthalein indicator solution	YAMAMOTO (1936)
	C	6.8 ± 0.2		

		pH	Method	Author(s)
Roussette (dogfish)				
egg cell	C	5.6—6.0	Crushing cells in indicator solution	REISS (1926 a)
Salmonidae:			Neutral red solution	TRIFONOVA (1934, 1937)
at fertilization	C	6.8		
32—64 cell stage	C	6.0		
blastodisc	C	7.8		
b) Amphibia				
Rana temporaria:				
unfertilized egg	AH	7.0—7.21	Electrometric measurement of living cells with the hydrogen electrode	DORFMAN (1936—1938)
	VH	6.76—7.21		
early cleavage	AH	7.10—7.44		
	VH	6.85—7.51		
morula	AH	7.39—8.08		
	VH	6.98—7.30		
blastula	AH	6.85—8.38		
	VH	6.77—7.88		
	DE	7.51—8.14		
early gastrula	VH	6.41 8.00		
	AH	7.53—8.39		
	DE	6.81—7.48		
late gastrula	AH	7.32—8.48		
	VH	7.67—8.00		
early neurula	AH	7.61—7.39		
tail bud stage	AH	6.77—7.39		
unfertilized egg	C	6.0	Crushing cells in indicator solution	REISS (1926 a)
Triton taeniatus:				
Egg cells in ovary	T	7.2	Antimony electrode	BUYTENDIJK and WOERDEMANN (1927)
	C	8.5		
2 blastomere stage	C	8.5		
4 blastomere stage	C	8.05		
32 blastomere stage	C	7.9		
64 blastomere stage	C	7.9		
Triton taeniatus:		early stages same as for *Rana temporaria*		
Blastula:				
outer cells	C	7.6—7.8	Antimony electrode	BUYTENDIJK and WOERDEMANN (1927)
blastocoel fluid		8.4—8.6		

		pH	Method	Author(s)
Triton taeniatus				
Gastrula:				
ectoderm	T	7.6—7.8		
endoderm	T	8.1		
Neurula:				
ectoderm	T	6.9—7.0		
neural tube	T	6.8		
early intestine	T	8.1		
yolk		6.9—7.0		
Rana (species not given)				
Frog muscle			Microinjection of	SCHMIDTMANN
Normal	MS	6.6—6.7	indicator particles	(1925 a, b)
Summer frog			Colorimetric pH	MEYERHOF and
Resting	MS	7.3	method on crushed cells and tissues	LOHMANN (1926)
Tetanized	MS	6.2—6.3		
Chloroformed	MS	5.8—6.0		
Winter frog, fasting			Microinjection of	
Resting	MS	7.3	Clark indicator	
Tetanized	MS	6.8	solution	
Resting	MS	7.5		
During contraction	MS	8.0		MARGARIA (1932, 1934)
Intercellular fluid				
Normal		7.4	Micro sample of intercellular fluid	FENN and MAURER (1935)
Injured		6.2		
2 hours after death	MS	6.8—7.0	Calculation by Henderson-Hasselbalch equation	
Muscle protoplasm		6.9		
Normal	MS	5.9—6.17	Crushing cells in indicator solution	REISS and VELLINGER (1926)
Tetanized	MS	6.0—6.06		
Asphyxiated	MS	5.87		
30 min. after death	MS	5.78		
Rana pipiens:				
Isolated muscle				
fibers			Microinjection of	WIERCINSKI (1952)
in 0.2 M $NaHCO_3$	C	8.4	Clark indicator	
in 0.12 M NaCl	C	6.0	solution into	
in mineral oil	C	5.8—6.4	isolated muscle	
in 0.186 M NH_4Cl	C	5.6	fiber with open (cut) ends	
Rana esculenta				
histiocytes (Summer)	C	6.4—7.2	Vital staining with acid pH indicators. Ameboid movement with India ink phagocytosis	HATANO and IWATA (1934)

		pH	Method	Author(s)
Necturus	N	7.5 ± 0.1	Microinjection of indicator solutions	CHAMBERS, POLLACK, and HILLER (1927)
Rana	N	7.5 ± 0.1		
c) Aves-Pigeon				
Columba spp.				
Flight	MS	6.1—6.8	Colorimetric method an mashed tissues and body fluids	GRÄFF and RAPPOPORT (1927)
Leg	MS	6.7—7.4		
Flight	MS	5.46—5.95	Electrometric measure of crushed cells and tissues	MCCARRISON, SANKARAN, and MADHAVA (1933)
Leg	MS	5.53—6.12		
Normal				
after 00 hours	MS	5.9—6.4		FOMIN and STRASHEKO (1934)
after 24 hours	MS	6.2—6.8		
after 48 hours	MS	7.8		
Avitaminosis B_1				
after 00 hours	MS	6.1—7.6		
after 24 hours	MS	6.9—7.8		
after 48 hours	MS	7.8		
Avitaminosis B_1				
Flight	MS	6.77—7.20		MCCARRISON, SANKARAN, and MADHAVA (1933)
Leg	MS	6.02—7.10		
d) Mammalia				See LISON (1941), GRÄFF and RAPPOPORT (1937) for other data for the Mammalia
Human blood:				
Arterial	S	7.400	Calculation by Henderson-Hasselbalch equation	DILL (1928), DILL, EDWARDS, and CONSOLAZIO (1927)
	E	7.190		
Venous	S	7.368		
	E	7.167		
Venous oxalated plasma		7.28—7.41	Colorimetric determination of Cullen corrected to electrometric pH at 38° C. for 27 subjects	Included in ALBRITTON (1952), GRAM (1924)
average		7.36		
Male	E	7.21—7.26		
average		7.24		

		pH	Method	Author(s)
Mus musculus				
White Mouse:				
Histiocytes	C	5.2—6.8	Vital staining with acid pH indicator dyes, India ink phagocytosis	Hatana and Iwata (1933, 1934)
Muscle fibers	T	6.0	Microinjection of Clark indicator solutions	Vlès and de Coulon (1924, 1925, 1928)
Blood plasma				
right heart		7.38	Injection of indicators into the blood stream. Examination of tissues under oil	Rous (1924—1926)
connective	T	7.2		
tendons	T	7.0		
Kidney cortex	T	5.0—6.6		
Macrophage	G	4.0		
Bone and cartilage	T	6.5		
Epidermis	T	7.0		
Lymph node	IS	6.2		
Spleen	IS	5.0		
Adipose				
liver spleen gall bladder heart muscle voluntary muscle pancreas	T	more 5.0 less 6.6		
Normal muscle	T	6.0—6.44	Crushing cells in indicator solution	Reiss and Vellinger (1926)
Tetanized muscle	T	5.9—5.95	Electrometric measure of living cells with hydrogen electrode	Reiss and Vellinger (1926)
30 minutes after death	T	5.8		
Normal mammary gland	T	6.4—7.0	Microinjection of Clark indicator solution	Chambers and Ludford (1932)
Mammary carcinoma	C	6.4—7.0	Microinjection of indicator solutions into tissue cultures of cells	Chambers and Ludford (1932)
Crocker carcinoma	N	7.2		
Tar carcinoma				

		pH	Method	Author(s)
Rattus rattus				
Flexner Rat:				
carcinoma	T	6.8	Glass electrode	MILLET (1928)
Liver	T	7.1	Tissues frozen in liquid air and pH determined at 18° C.	
Spleen	T	7.1		
Cardiac muscle				
Guinea pig *Cavia porcellus*	MS	6.8—7.0	Electrometric measure of tissue frozen in liquid air	MICHAELIS and KRAMSZTYK (1914)
Cat *(Felis catus)*	MS	6.8—7.0		
Dog *(Canis familiaris)*	MS	6.4—7.0		
Dog	MS	6.8	Microinjection of Clark indicator particles	SCHMIDTMANN (1924—1927)
Cartilage				
Chondrocytes				
Guinea pig, Cat:	T	6.6—6.8		
bronchial	T	6.9—7.1		
Rat	T	6.5—7.2	Injection of indicators into the blood stream, examination of tissues under oil	ROUS (1925 c)
Digestive tract Cat:				
Marginal cells	C	6.4—6.6	Microinjection of Clark indicator particles	SCHMIDTMANN (1924, 1927)
Chief cells	C	6.6—7.0		
Duodenum	C	6.8—7.0		
Jejunum	C	7.2—7.3		
Ileum and Large intestine	C	7.4—7.6		
Guinea pig and rabbit:				
Peripheral cells	C	7.1—7.5		
Central cells	C	6.7—7.0		
Kupfer cells	C	6.4—6.5		
	T	6.0—6.5	Electrometric measure of crushed cells and tissues	DUVAL, GUEYLARD, and PORTIER (1925)
Pancreas Cat, guinea pig and rabbit:				
Parenchyma	C	7.2—7.4	Microinjection of Clark indicator particles	SCHMIDTMANN (1924—1927)
Islet	C	6.6—6.9		

		pH	Method	Author(s)
Kidney				
Cat	T	6.98	Electrometric measure of tissue frozen in liquid air	MICHAELIS and KRAMSZTYK (1914)
Various mammals				
Loop of Henle	C	6.7—7.4	Microinjection of Clark indicator particles	SCHMIDTMANN (1924, 1927)
convoluted tubules	C	6.7—7.4		
reticulate cells	C	6.4—6.6		
Ovary				
Rat and Cat:				
egg	C	7.5—7.8		
Follicular cells	C	7.0—7.3		
Conjunctival cells	C	6.7—6.9		
White rat and mouse:				
Histiocytes	C	6.4—6.8	Vital staining with acid pH indicators and India ink phagocytosis	HATANO and IWATA (1933, 1934)
White Rat:				
Testicle				
Spermatogonia cortical cells	C	7.4	Microinjection of Clark indicator particles	SCHMIDTMANN (1927)
inner cells	C	6.8		
spermatozooid	C	7.4		
interstitial cells	C	6.6—6.8		
Cat:				
Epithelium				
surface	T	7.3—7.5		
base	T	6.8—6.9		
horny layer	T	6.2		
Rat:				
Uterus				
pseudopregnant	T	acid	Perfusion of tissue (buffering capacity)	HALL (1936)
Uterus				
Rabbit and Cat:				
non-pregnant	T	6.8 less	Perfusion of tissues (buffering capacity)	NAGAKAWA (1935)
involuting	T	6.6 more		TIFFENEAU and BROUN (1935)

		pH	Method	Author(s)
Horn of Uterus Cat:				
Anesthesia and bleeding	T	7.07 ± 0.07	Perfusion of tissues (buffering capacity)	KERRIDGE and WINTON (1929)
Minced	T	6.22—7.03		
Average		6.58		
Guinea pig	T	6.94—6.96		
Striated muscle Rat	MS	7.4—7.7	Glass electrode on 1 to 2 grams of minced muscle, moistened with M/22 sodium iodoacetate	BATE-SMITH (1948)
Whale				
Balaenoptera musculus (Blue)	MS	7.60 ± 0.08	Glass electrode on 1 to 2 grams of minced muscle in 10 ml. of 0.005 M sodium acetate	MARSH (1952)
physalus (Fin)	MS	7.44 ± 0.13	Glass electrode on 1 to 2 grams of minced muscle in 10 ml. of 0.005 M sodium acetate	MARSH (1952)
Physeter catadon (Sperm)	MS	7.68 ± 0.07		
Human				
Neoplasms Innocent	MS	6.58	Glass electrode. Tissues frozen in liquid air and the pH determined at 18° C.	MILLET (1928)

VI. Discussion

The definition of pH, its application to the problem of pH in protoplasm, the methods used in the determination of the pH of cells and tissues, and the results obtained have been outlined in earlier sections. Obviously, the results obtained by various workers are very different and the conclusion is inescapable that the differences are in many cases due to the method used.

GRÄFF and RAPPOPORT (1937), BROOKS (1941), LISON (1941), BRACHET (1944, 1950), GUILBERT and LANGLAIS (1951) and HEILBRUNN (1952) discuss the limitations and give criticisms in relation to the work on various kinds of cells and pH.

A. Nucleus

Different techniques used for the determination of nuclear pH in the living cell have yielded different values.

CROZIER (1918, 1923) using natural indicators found a pH of 6.7 for the nuclei of *Opalina* (Infusoria). SCHMIDTMANN (1924, 1925 a) reported pH 6.7 for the ectodermal cell nuclei of *Hydra.* Her value is based on the micro-injection of indicator particles. REISS (1924 a, b, 1928 a) obtained a pH of 7.2–7.5 for the nucleus in cells of *Sabellaria alveolata* and *Arenicola* (*Annelida*) by crushing these cells in indicator solution.

REISS (1924 b) used the microcolorimetric method on the nucleus of *Echinocardium, Sabellaria, Psammechinus,* and *Sipunculus.* It was found that the nucleus can vary in pH from less than 5 to 9. Anesthetics can raise the pH somewhat. Also, see CLOWES, KELTCH, and KRAHL (1940).

RAPKINE and WURMSER (1926) by means of microinjection of indicator dyes found the nuclei of Echinoderm eggs and the salivary gland cells of *Chironomus* to be in the range of 7.0—7.3. NASSANOV (1932) used vital staining in 0.25% concentration in various Infusoria. He concluded that the pH of the macronucleus was about 6.0.

CHAMBERS and LUDFORD (1932) microinjected indicator solutions into tissue cultures of cells of Crocker carcinoma (mammary gland of the white mouse). They found a pH of 7.2 for the nucleus.

BERTANI (1944) injected Clark and Lubs indicators into the salivary gland cell of *Chironomus* larvae. He found a cytoplasmic pH in the vicinity of 7.0. For the chromosomal pH in the nucleus the indicators showed a pH between 5.8–6.2.

CHAMBERS (1949) in reviewing work of his own states the nuclear pH to be in the range 7.6 to 7.8.

B. Cytoplasm

VLÈS (1926) and POLLACK (1928) state that too much phenol red was injected by NEEDHAM and NEEDHAM (1925 a, b), thereby masking the effect of the true color of the protoplasm in ameba. Cytolysis took place in two or three minutes after the organism were injected and the animals were considered not to be normal.

RAPKINE and BOUXIN (1926) state the internal pH of the larvae of *Paracentrotus lividus* during the formation of calcareous spicules was in the range from 7.3 to 8.5 pH. When the external pH was lowered from the pH of sea water 8.1 to a value such as 5.4, with sulfuric acid, cytolysis of the larvae occurred.

CHAMBERS, POLLACK, and HILLER (1927) conclude that the living protoplasm of widely differing types of cells, i. e., Echinoderm ova, amebae, tissue and germinal cells of Necturus and the frog under normal conditions apparently have a remarkably constant pH value which is slightly on the acid side, while the nuclear pH is distinctly on the alkaline side of neutrality. Injury produces an increased acid reaction in the cytoplasm but does not affect the pH of the nucleus.

Reiss (1928 a, b) with the technique of crushing for the cytoplasm of the *Paracentrotus lividus* egg concludes a value of 5.6.

Vlès and Vellinger (1928) find a value of 5.5 ± 0.3 with spectrophotometric technique.

Chambers (1929) and his co-workers, with microinjection, arrive at a pH of 6.8 ± 0.2 for the cytoplasm of all of the cells that they examined.

Chambers, Cohen and Pollack (1931) immersed Echinoderm ova in indicators and dyes dissolved in sea water. Of the simple and substituted indophenols used, all except those containing a sulfonated radical could be detected in living Echinoderm ova. All the indophenols containing the sulfonate radical and the indigo sulfonates do not penetrate. The amphoteric dyes and the basic dyes readily penetrate the ova.

Chambers and Ludford (1932) claim that tumor cells of mammals are in accordance with normal cells.

Chambers (1932) compares the value of 6.8 ± 0.1 for echinoderm eggs, obtained by microinjection of indicators with that of Bodine (1927) who used the electrometric method for the egg of the fish *Fundulus* and obtained a value of 6.38. The electrometric method thus gives values which are on the acid side of those obtained by the colorimetric method; but, of course, the interior of the fish egg consists of yolk, not living protoplasm. There is injury produced by both. Chambers, by the method of microinjection of indicators, obtained the value of 6.8 ± for the cytoplasm of the *Fundulus* egg. His acid of injury value is 5.4 ± as for other cells.

Spek and Chambers (1933) worked with *Ameba*. They assigned a value of pH 7.3 for the hyaloplasm microinjected with sulfonphthalein pH indicators. This is higher than the values that Chambers (1928) and Cohen, Chambers, and Reznikoff (1928) found.

Yosezato (1933) used fifteen indicators on *Entamoeba histolytica*. He immersed the organism in a solution of the indicator. The organism was obtained from the faeces of man and the kitten. The pH of the cytoplasm was very close to 7.0.

Trifonova (1934) used neutral red to determine pH localization in fish eggs. Trifonova (1937) followed changes in pH during the development of fish eggs. At fertilization the pH of the egg is 6.8, during development in the 32–64 blastomere stage the pH is 6.0, and the blastodisc pH is 7.8.

Trifonova, Vernidoube, and Philippov (1939) working with salmon eggs show that the quantity of lactic acid produced in the first 18 hours of development after fertilization of the egg progresses from 3.0 mgm./hr. at the 3rd hour after fertilization to 0.2 mgm./hr. after 18 hours.

Reiss and Gersch (1936) with vital dyes claim the macromeres and micromeres of *Aplysia limacina* have a pH ranging between 6.0 and 7.0.

Yamamoto (1936) working with *Oryzias* eggs (Pisces) injected sulfonphthalein indicators. To estimate the pH of the yolk, eggs were fertilized and used the same day. For pH of the cytoplasm, eggs in the 1–4 cell stages were used. The pH of the yolk at this stage is 6.4 ± 0.2 which comes very close to the figure obtained potentiometrically by Bodine (1927) with the *Fundulus* egg. The pH of the cytoplasm, is about 6.8 ± 0.2 which coincides with Chambers' (1932) colorimetric study of the cytoplasm of *Fundulus* eggs.

Matsuda (1936 a, b) used 13 different indicators on T. vaginalis. The technique was the same as that of Yosezato (1933) and the organisms came from vaginal secretion and from Tanabe-Chiba's medium. The cytoplasmic pH ranged from 5.5 to 6.8.

Grand (1938) used cresol red, phenol red, bromcresol purple, chlorphenol red, and bromcresol green on egg cells of *Mactra solidissima* (Mollusca). None of these indicators served as vital stains in ordinary immersion methods except when the cells were dead. Dead cells stain intensively with most of these indicators, the pH reaction of the cytolyzed material being the same as that of sea water. This fact made it possible to determine those cases in which the microinjections were performed without injury. The eggs were operated upon while immersed either in normal alkaline (pH 8.4) or acidified (pH 5.6) sea water. This procedure was used to check against the possibility that cytoplasmic pH was influenced by the surrounding sea water. Grand, a co-worker of Chambers found a pH value of 6.8.

Spek (1940) states that the stain base shows a spectrum resembling that of the metachromatic solution. The de-ionizing of the salt is the mechanism producing the change of color.

Although Chambers stoutly maintains that the pH of the cytoplasm of all living cells at all times is 6.8 ± 0.2, this can scarcely be true unless one completely discards the results of most other investigators. Moreover, the work of Runnström (1949) and Borei (1948, 1949) clearly shows that following fertilization the protoplasm of the sea urchin egg exhibits a marked increase in acidity. This fact is obviously in conflict with the idea that the pH of all cytoplasm is always the same.

C. Vacuolar pH

Strelnikov (1929) found that some basic dyes were obsorbed in *Paramecium* more than 25,000 times that found in the external medium.

Rumjantzew and Kedrowsky (1926) found that vital staining with basic dyes was fixed in the granules and vacuoles of *Opalina.* They also crushed and ground *Opalina,* and added indicator to the extract. They obtained a pH of 6.1 to 6.4. Crozier placed *Paramecium* and *Opalina* in solutions of acidic and basic indicators until the dye penetrated. He concluded that the reaction of the protoplasm was acid.

Raven (1937) with vital dyes claims the cytoplasmic inclusions of *Chaetopterus, Nereis* and *Aplysia* eggs are acid and basic. The hyaline zone was not colored by the indicators.

Spek (1934 a, 1937) with vital dyes claims the granules and vacuoles as protoplasmic constituents vary from pH 5 to 8.

In conclusion, it must be recognized that in the study of the hydrogen ion concentration of various types of protoplasm, most investigators have not been careful in considering possible errors involved in the methods they used. Perhaps some of the differences in the values obtained may be due to such errors. However, no one surveying the literature critically could well agree that the generalization is correct in maintaining that all cells at all times have the same hydrogen ion concentration.

Bibliography

ALBERT, A., 1952: Ionization, pH and biological activity. Pharmacol. Rev. **4**, 136—167.

ALBRITTON, E. C., 1952: Standard Values in Blood. Philadelphia.

ALEXANDROV, W., 1932: Über die Bedeutung der Oxydo-Reduktion, Bedingungen für die vitale Färbung, mit besonderer Berücksichtigung der Kernfärbung in lebendigen Zellen. Protoplasma **17**, 161—217.

ARMSTRONG, P. B., 1927: Determination of the pH of developing *Fundulus* eggs. Proc. Soc. exper. Biol. a. Med. (Am.) **25**, 146—147.

ARRHENIUS, S., 1887 a: Einfluß der Neutralsalze auf die Reaktionsgeschwindigkeit der Verseifung von Äthylacetat. Z. physik. Chem. **1**, 110—133.

— 1887 b: Über die Dissociation der in Wasser gelösten Stoffe. Z. physik. Chem. **1**, 631—648.

— 1889: Über die Reaktionsgeschwindigkeit bei der Inversion von Rohrzucker durch Säuren. Z. physik. Chem. **4**, 226—248.

ARVANITAKI, A., and N. CHALAZONITIS, 1951: Recherches sur la répartition de quelques catalyseurs respiratoires dans l'espace cellulaire (axon géant et soma neuronique de Sepia). Arch. sci. physiol. **5**, 207—226.

ATKINS, W. R. G., 1922: Dibromthymolsulphonephthalein as a reagent for determining the hydrogen ion concentration of living cells. J. Mar. Biol. Assoc. (United Kingdom). **12**, 781—784.

BATE-SMITH, E. C., 1948: On the observation of pH and the related properties of meat. J. Soc. Chem. Ind. **67**, 83—90.

BENEDETTI-PICHLER, A. A., 1937: Qualitative analysis of microgram samples. Ind. Eng. Chem. Anal. Ed. **9**, 483—487.

— and J. R. RACHELE, 1940: Limits of identification of simple confirmatory tests. Ind. Eng. Chem. Anal. Ed. **12**, 233—241.

BENSON, C. C., 1928: Hydrogen ion concentration of fish muscle. J. biol. Chem. **78**, 583—590.

BERNAL, J. D., and R. H. FOWLER, 1933: A theory of water and ionic solution, with particular reference to hydrogen and hydroxyl ions. J. Chem. Phys. **1**, 515—548.

BERTANI, G., 1944: Contributo allo studio del pH intranucleare nelle cellule delle ghiandole salivari di *Chironomus thummi*. Rend. ist. lombardo sci. ser. 3, **77**, 427—436.

BJERRUM, N., 1917: Die moderne Auffassung der sauren und basischen Reaktion und ihre Anwendung in der Analyse. Benützte Bezeichnungen. Z. anal. Chem. **56**, 13—28; Die Farbstoffindikatoren. **56**, 81—95.

BOCHE, R. D., and J. B. BUCK, 1942: Studies on the hydrogen ion concentration of blood and their direct bearing on in-vitro cytological technique. Physiol. Zool. **15**, 293—303.

BODINE, J. H., and D. E. FINK, 1925: A simple micro vessel with electrode for determining the hydrogen ion concentration of small amounts of fluid. J. gen. Physiol. **7**, 735—740.

— 1927: Potentiometric studies on intracellular pH values of single *Fundulus* egg cells. J. gen. Physiol. **10**, 533—540.

BOREI, H., 1948: Respiration of oocytes, unfertilized eggs and fertilized eggs from *Psammechinus* and *Asterias*. Biol. Bull. **95**, 124—150.

— 1949: Independence of post-fertilization respiration in the sea urchin egg from the level of respiration before fertilization. Biol. Bull. **96**, 117—122.

BOURNE, G. H., 1951: Cytology and Cell Physiology. 2nd Edition. London.

BOZLER, E., 1924: Über die Morphologie der Ernährungsorganellen und die Physiologie der Nahrungsaufnahme von *Paramecium caudatum*. Arch. Protistenk. **49**, 163—215.

BRACHET, J., 1944: Embryologie Chimique. Paris.
— 1950: Chemical Embryology. English translation by L. G. Barth. New York.
BRITTON, H. T. S., 1942: Hydrogen ions. 3rd Edition. London.
BROCKMAN, C. J., 1935: A symposium on indicators. Chem. Rev. **16**, 53—55.
BROOKS, S. C., and M. M. BROOKS, 1941: The permeability of living cells. Berlin. Reprinted 1944. Ann Arbor.
BROWN, J. H., 1923: The colorimetric determination of the hydrogen ion concentration of small amounts of fluid. J. Labor. a. clin. Med. (Am.) **9**, 239—244.
BUYTENDIJK, F. J., and M. W. WOERDEMANN, 1927: Die Messung der Wasserstoffionenkonzentration. Arch. Entw.mechan. **2**, 387—410.
CARNOT, P., R. GLÉNARD, et Mme. GRUZEWSKA, 1925: Les colorations vitales au rouge neutre comme indices de la concentration ioniques des organes vivants. C. r. Soc. Biol. **92**, 865—868.
CALDWELL, P. C., 1953: Measurement of intracellular pH by means of a tungsten micro-electrode. J. Physiol. **120**, 31 P.
— 1954: An investigation of the intracellular pH of crab muscle fibres by means of micro-glass and micro-tungsten electrodes. J. Physiol. **126**, 169—180.
— and E. J. HARRIS, 1952: Metabolic control by intracellular pH. Biochem. J. **51**, xli.
CHAMBERS, R., 1922: New apparatus and methods for the dissection and injection of living cells. Anat. Rec. (Am.) **24**, 1—19.
— 1928: The relation of the environment to the pH of protoplasm and of its inclusion bodies. Biol. Bull. **55**, 369—376.
— 1929 a: Vital staining with methyl red. Proc. Soc. exper. Biol. a. Med. (Am.) **27**, 809—811.
— 1929 b: Hydrogen ion concentration of protoplasm. Bull. National Research Council (U. S.) **69**, 37—47.
— 1932: The pH of the protoplasm of the *Fundulus* egg. J. cellul. a. comp. Physiol. (Am.) **1**, 65—70.
— 1940: Micromanipulation of the living cell. Amer. Assoc. Advance. Sci. Publication **14**, 20—30.
— 1943: Electrolytic solutions compatible with the maintenance of protoplasmic structures. Biol. Symposia. **10**, 91—109.
— 1949: Micrurgical studies on protoplasm. Biol. Rev. **24**, 246—265.
— B. COHEN, and H. POLLACK, 1931: Permeability of Echinoderm ova to indicators. J. exper. Biol. **8**, 1—8.
— — — 1932: Reduction potentials of European marine ova and *Amoeba proteus* as shown by indicators. Protoplasma **17**, 376—387.
— and H. POLLACK, 1926: Colorimetric determination of the nuclear and cytoplasmic pH in the starfish egg. J. gen. Physiol. (Am.) **10**, 739—755.
— H. POLLACK, and B. COHEN, 1929: Reduction potentials of marine ova as shown by indicators. Brit. J. exper. Biol. **6**, 229—247.
— and R. J. LUDFORD, 1932: Colorimetric pH of malignant cells in tissue culture. Proc. roy. Soc. **110 B**, 120—124.
— H. POLLACK, and S. HILLER, 1927: The protoplasmic pH of living cells. Proc. Soc. exper. Biol. a. Med. (Am.) **24**, 760—761.
COHEN, B., R. CHAMBERS, and P. REZNIKOFF, 1928: Reduction potential of *Amoeba dubia* by microinjection of indicators. J. gen. Physiol. (Am.) **11**, 585—611.
CLARK, A. J., 1933: The mode of action of drugs on cells. London.
CLARK, W. M., 1928: The determination of hydrogen ions. 3rd Edition. Baltimore.
CLOWES, G. H., A. K. KELTCH, and M. E. KRAHL, 1940: The role of changes in extracellular and intracellular hydrogen ion concentration in the action of local anaesthetic bases. J. Pharmacol. (Am.) **68**, 312—329.
CONWAY, E. J., 1947: Exchanges of potassium, sodium and hydrogen ions between cell and environment. Ir. J. med. Sci. 6th Series, 654—680.
— and P. J. FEARON, 1944: Acid-labile carbon dioxide in mammalian muscle and hydrogen ion concentration of the muscle fiber. Nature **153**, 54—55.
CORSON, S. A., 1943: A quantitative method of injecting controlled quantities of aqueous solutions into living cells. Proc. Oklahoma Acad. Sci. **23**, 31—32.
COWAN, S. L., 1933: The CO_2 dissociation curves and the buffering of crab's muscle and nerve preparations. J. exper. Biol. **10**, 401—411.
CROZIER, W. J., 1918: On indicators in animal tissues. J. biol. Chem. (Am.) **35**, 455—460.
— 1923: A note on the reaction of protoplasm. Proc. Soc. exper. Biol. a. Med. (Am.) **21**, 58.

DANIELLI, J. F., 1937: Relations between surface pH, ion concentrations and interfacial tension. Proc. roy. Soc. **B 122**, 155—174.

— 1941: On the pH at the surface of ovalbumin molecules and the protein error with indicators. Biochem. J. (Brit.) **35**, 470—478.

— 1946: A critical study of techniques for determining the cytological position of alkaline phosphatase. J. exper. Biol. **22**, 110—117.

DANIELS, F., J. H. MATHEWS, and J. W. WILLIAMS, 1941: Experimental Physical Chemistry. New York.

DATE, S., 1931: Concentration en ions hydrogène dans un corps unicellulaire. C. r. Soc. Biol. **106**, 89—93.

DILL, D. B., 1928: The calculation of cell volume changes as a function of hydrogen ion concentration. J. biol. Chem. (Am.) **76**, 543—545.

— H. T. EDWARDS, and W. V. CONSOLAZIO, 1937: Blood as a physicochemical system. XI. Man at rest. J. biol. Chem. (Am.) **118**, 635—648.

DISTECHÉ, A., and M. DUBUISSON, 1954: Transient response of the glass electrode to pH step variation. Rev. Sci. Instr. **25**, 869—875.

DOLE, M., 1941: The Glass Electrode. New York.

DORFMANN, W. A., 1936: A simple type of microelectrode for the determination of pH and Eh. Protoplasma **25**, 465.

— 1938: The pH and anaerobic redox potential in early Amphibian morphogenesis. Bull. biol. med. exper. U. R. S. S. **6**, 413—417.

— and D. E. GRODENSKY, 1937: The oxidation-reduction and pH gradient of the unfertilized Amphibian egg and the technique of determination of these values in situ. Bull. biol. med. exper. U. R. S. S. **4**, 265—268.

DRURY, D. R., and P. ROUS, 1926 a: Influence of lymph soluble tissue materials on the significance of the coloration with some phthalein indicators. J. exper. Med. (Am.) **43**, 669—686.

— — 1926 b: Influence of lymph insoluble tissue materials on the significance of the coloration with some phthalein indicators. J. exper. Med. (Am.) **43**, 687—701.

DUBUISSON, M., 1936: pH changes of muscle during and after contraction. Proc. Soc. exper. Biol. a. Med. (Am.) **35**, 609—611.

— 1939: Studies on the chemical processes which occur in muscle before, during and after contraction. J. Physiol. (Brit.) **9 h**, 461—482.

DUVAL, M., F. GUEYLARD, et P. PORTIER, 1925: Réaction ioniques du foie dans la série animale. C. r. Soc. Biol. **92**, 484.

EASTMAN, E. D., and G. K. ROLLEFSON, 1947: Physical Chemistry. 1st Edition. New York.

EPHRUSSI, B., et A. NEUKOMM, 1927: Sur la variations de l'inhibition de l'œuf d'oursin en fonction du pH de l'eau de mer. C. r. Soc. Biol. **96**, 1196—1198.

FAURÉ-FRÉMIET, E., 1923: Variations de l'alcalinité de l'œuf de *Sabellaria* pendant la maturation. C. r. Soc. Biol. **88**, 863—866.

FENN, W. O., 1928: The carbon dioxide dissociation curve of nerve and muscle. Amer. J. Physiol. **85**, 207—223.

FENN, W. O., and F. W. MAURER, 1935: The pH of muscle. Protoplasma **24**, 337—345.

FINK, D. E., 1925: Metabolism during embryonic and metamorphic development of insects. J. gen. Physiol. (Am.) **7**, 527—543.

FOLLOWS, A. B., 1934: A study of mistakes in colour matching made by persons with defective color vision. J. Physiol. (Brit.) **82**, 172—178.

FOMIN, S. V., et D. N. STRAJEKO, 1934: Étude des constantes physico-chimiques du tissu musculaire. Conductibilité électrique et le pH dans l'autolyse du tissu musculaire chez les pigeons alleints d'avitaminose B. Ukrain. Biochem. J. **7**. 117—124.

GERSCH, M., 1936: Untersuchungen über vitale Kern- und Plasmafärbung bei *Paramecium*. Ver. dtsch. Zool. ges. Anz. **9**, 268—274.

— 1937: Vitalfärbung als Mittel zur Analyse physiologischer Prozesse (Untersuchungen an *Paramaecium caudatum*). Protoplasma **27**, 412—441.

GILLESPIE, L. J., 1920: Colorimetric determination of titration curves without buffer mixtures. J. amer. chem. Soc. **42**, 742—749.

GLASER, R. W., 1925: Hydrogen ion concentration in the blood of insects. J. gen. Physiol. (Am.) **7**, 599—602.

GLASSTONE, S., 1946: Textbook of Physical Chemistry. New York.

GORTNER, R. A., 1930: The state of water in colloidal and living systems. Trans. Faraday Soc. **26**, 678—686.

GRÄFF, S., 1924: Ein Verfahren zur Bestimmung der Wasserstoffionenkonzentration im Gewebe mit Indikatoren. Klin. Wochenschr. **3**, Nr. 11.

— und E. RAPPOPORT, 1937: Methoden und Ergebnisse der Bestimmung der Wasserstoffionenkonzentration des tierischen Gewebes. Erg. Path. und Path. Anat. **33**, 181—279.

GRAM, H. C., 1924: Composition and physical properties of normal human blood: A compilation of curves from the literature. Amer. J. med. Sci. **168**, 511—526.

GRAND, C. G., 1938: Intracellular pH studies on the ova of *Mactra solidissima.* Biol. Bull. (Am.) **75**, 369.

GUILBERT, C., and L. LANGLAIS, 1951: pH et rH en biologie. Sem. Hôp. Par. **27**, 424—430.

HALL, B. V., 1936: Variations in acidity and oxidation-reduction potential of rodent uterine fluids. Physiol. Zool. **9**, 471—497.

HARTLEY, G. S., and J. W. ROE, 1940: Ion concentrations at interfaces. Trans. Faraday Soc. **36**, 101—109.

HATANO, S., and S. IWATA, 1933: Hydrogen ion concentration of histiocyte by vital staining with indicator dyes. J. orient. Med. (Mandsch.) **19**, 67.

— — 1934: Hydrogen ion concentration of histiocyte by vital staining with indicator dyes. J. orient. Med. (Mandsch.) **20**, 35.

HEILBRUNN, L. V., 1952: An Outline of General Physiology. 3rd Edition. Philadelphia.

HILL, D. K., 1940: Hydrogen ion concentration changes in frog's muscle following activity. J. Physiol. (Brit.) **98**, 467—469.

HOET, J. P., and P. M. T. KERRIDGE, 1926: Observations on the muscles of normal and moulting Crustacea. Proc. roy. Soc., Lond. **100 B**, 116—119.

HOMER, A., 1917: A note on the use of indicators for the colorimetric determination of the hydrogen ion concentration of sera. Biochem. J. **11**, 283—291.

HOWLAND, R. B., 1928: The pH of gastric vacuoles. Protoplasma **5**, 127—134.

— 1930: The pH of the vacuolar fluid in *Actinosphaerium* eichhorni. J. exper. Zool. **55**, 53—62.

JACOBS, M. H., 1920: The production of intracellular acidity by neutral and alkaline solutions containing carbon dioxide. Amer. J. Physiol. **53**, 547—463.

JAUMAIN, D., 1925: Importance de l'erreur de protéine dans la détermination du pH à l'aide du bleu de bromthymol. C. r. Soc. Biol. **93**, 860—862.

KAMADA, T., 1935: Contractile vacuole of *Paramecium.* J. Fac. Sci. Imp. Univ. Tokyo **4**, 49—61.

KEDROWSKY, B., 1931: Methodik der Kultur in künstlichen Medien, pH-Regulierungen und Ionen-Gleichgewichte im Kultur-Medium. Protoplasma **12**, 356—379.

KERRIDGE, P. T., and F. R. WINTON, 1929: The hydrogen ion concentration of the isolated uterus. J. Physiol. (Brit.) **67**, 66—76.

KILPATRICK, M., 1935: The colorimetric determination of hydrogen ion concentration in aqueous solutions. Chem. Rev. **16**, 57—66.

— and M. L. KILPATRICK, 1932: The teaching of the theory of the dissociation of electrolytes. 2. The definition of pH. J. Chem. Ed. **9**, 1010—1016.

KITE, G. L., 1913: Studies on the physical properties of protoplasm. 1. The physical properties of certain animal and plant cells. Amer. J. Physiol. **32**, 146—164.

KOLTHOFF, I. M., 1937: Acid-base indicators. Translated by C. Rosenblum. New York.

— and N. H. FURMAN, 1926: Indicators. New York.

KOPAC, M. J., 1935: Intracellular pH determinations on marine ova. Ann. Rept. Tortugas Lab., Carnegie Inst. Wash. Year Book **34**, 85—86.

KOPACZEWSKI, W., 1928: Pénétration électro-capillaire des matières colorantes dans la cellule. C. r. acad. sci. **186**, 1758—1761.

LEE, J. W., 1942: The effect of pH on food vacuole formation in *Paramecium.* Physiol. Zool. **15**, 459—465.

LISON, L., 1935: Études sur la métachromasie. Colorants métachromatiques et substances chromotropes. Arch. Biol. (Fr.) **46**, 599—668.

— 1941: pH et rh intra-cellulaires. Tabulae Biologicae (The Hague) **19**, 1—23.

LUCKÉ, B., 1925: Observations in the intra-vital staining of centrifuged marine eggs. Proc. Soc. exper. Biol. a. Med. (Am.) **22**, 305—306.

MARGARIA, R., 1932: Metodo diretto di registrazione del musculo durante la contrazione. Boll. Soc. ital. Biol. sper. **7**, 557—561.
— 1934: An apparent change of pH on stretching a muscle. J. Physiol. (Brit.) **82**, 496—497.
— e C. PULCHER, 1934: Dimostrazione dell' alcalinizzazione del musculo durante la contrazione. Bull. Soc. ital. Biol. sper. **9**, 879—880.
MARSH, B. B., 1952: Observations on rigor mortis in whale muscle. Biochem. et Biophys. Acta **9**, 127—132.
MAST, S. O., 1947: The food vacuole in *Paramecium*. Biol. Bull. (Am.) **92**, 31—72.
MATSUDA, K., 1936 a: Studies on *Trichomonas vaginalis* Donne. J. orient. Med. (Mandsch.) **24**, 35—36.
— 1936 b: Hydrogen ion concentration of the cytoplasm of *T. vaginalis*. J. orient. Med. (Mandsch.) **24**, appendix 34.
— and Y. KOYAMA, 1949: Salt error of brom-thymol blue; note on colorimetric determinations of pH with brom-thymol blue at 38° C. Tôhoku J. exper. Med. (Jap.) **50**, 235—239.
McCARRISON, R., G. SANKÄRÄN, and K. B. MADHÄVÄ, 1933: Hydrogen ion concentration in organs of pigeons fed on polyneuritis producing diets. Indian J. Med. Res. **20**, 739—756.
McCUTCHEON, M., and B. LUCKÉ, 1924: The mechanism of vital staining with basic dyes. J. gen. Physiol. (Am.) **6**, 501—507.
MICHAELIS, L., 1947: The nature of the interaction of nucleic acids and nuclei with basic dyestuffs. Cold Spring Harbor Symposia Quant. Biol. **12**, 131—142.
— und A. KRAMSZTYK, 1914: Die Wasserstoffionenkonzentration der Gewebesäfte. Biochem. Z. **62**, 180—185.
MILLET, H., 1928: Measurement of the pH of normal, fetal, and neoplastic tissues by means of the glass electrode. J. biol. Chem. (Am.) **78**, 281—288.
MORITA, Y., and R. CHAMBERS, 1929: Permeability differences between nuclear and cytoplasmic surfaces in *Amoeba dubia*. Biol. Bull. (Am.) **56**, 64—67.
MURRAY, H. D., and C. A. SPENCE, 1939: Colour in Theory and Practice. London.
— 1952: Colour in Theory and Practice. London.
MEYERHOF, O., und K. LOHMANN, 1926: Über die Vorgänge bei der Muskelermüdung. Biochem. Z. **168**, 128—165.
NAGAKAWA, K., 1935: Über die Pufferungspotenz des Kaninchenuterus. Jap. J. med. Sci. trans. 4 pharmacol. **8**, 81—82.
NASSANOV, D., 1932: Vitalfärbung des Makronucleus aerober und anaerober Infusorien. Protoplasma **17**, 218—238.
NEEDHAM, J., and D. M. NEEDHAM, 1925 a: The hydrogen ion concentration and the oxidation-reduction potential of the cell interior: a microinjection study. Proc. roy. Soc., Lond. **99 B**, 259—286.
— — 1925 b: The hydrogen ion concentration and the oxidation-reduction potential of the cell interior. J. Physiol. (Brit.) **59**, 77.
— — 1926: The hydrogen ion concentration and the oxidation-reduction potential of the cell interior before and after fertilization and cleavage. A microinjection study on marine eggs. Proc. roy. Soc., Lond. **99 B**, 173—199.
NERENSTEIN, E., 1925: Über die Natur und Stärke der Säureabscheidung in den Nahrungsvakuolen von *Paramecium caudatum*. Z. wiss. Zool. **125**, 513—518.
NIMS, L. F., 1938: Glass electrodes and application for direct recording in vivo. Yale J. Biol. a. Med. (Am.) **10**, 241—246.
NOMURA, S., and S. KOKUBO, 1934: The hydrogen ion concentration of the living tissue in the hydromedusa, *Aglantha digitalis* Haeckel. Protoplasma **20**, 85—89.
— 1939: Intracellular oxidation-reduction potential limiting the ciliary movement. Protoplasma **20**, 85—89.
OGAWA, J., 1928: Examination of H ion concentration by means of micromanipulation. VI. Experiments on the reaction of tissues. Tokyo Imp. Univ. Central J. Med. **26**, 889—896.
— 1929: Studien über intrazelluläre Wasserstoffionenkonzentration der *Entamoeba histolytica* und *Entamoeba coli*. Zbl. Bakter. Parasitenk. **114**, 68—81.
OGSTON, A. G., 1947: The definition and meaning of pH. Physiol. Rev. (Am.) **27**, 228—239.
ONGARO, D., 1931 a: La concentratione idrogenionica nell'uovo del filugello. Annuar. staz. Bacol. sper. Padova. **XLVI**, 331—340.
— 1931 b: Influenca della fecondazione e dei trattamenti per l'accelerazione dello sviluppo embrionale. Annuar. staz. sper. Padova **XLVI, 341—350.**

Pandit, C. G., and R. Chambers, 1932: The pH of the egg of the sea urchin, *Arbacia punctulata*. J. cellul. a. comp. Physiol. (Am.) **2**, 243—349.
Pantin, C. F. A., 1923 a: On the physiology of ameboid movement. J. Mar. Biol. Assoc. (United Kingdom) **13**, 24—69.
— 1923 b: The determination of pH of microscopic bodies. Nature **111**, 81.
Parat, M., 1928: Contribution à l'étude morphologique et physiologique du cytoplasme. Troisième partie. Arch. Anat. microsc. **24**, 297—319.
Pastori, G., 1932: Variazioni del pH nei vacuoli delle Vorticelle in sequito a contrazioni provocate. Arch. Sci. biol. (It.) **17**, 164—168.
Peterfi, T., 1928: Ein Beitrag zur Methode der pH-Bestimmung in Zellen und Geweben. Z. Mikrosk. **45**, 56—59.
Peters, R. A., 1929: Co-ordinative biochemistry of the cell and tissues. "Cell Surfaces." J. State Med. **37**, 683—709.
Pfeiffer, H., 1927: Der gegenwärtige Stand der kolorimetrischen Azidimetrie in der Gewebephysiologie. Eine kritische Umschau unter umstrittenen Fragen. Protoplasma **1**, 434—465.
Pollack, H., 1928: Intracellular hydrion concentration studies. Biol. Bull. (Am.) **55**, 383.
Rapkine, L., et H. Bouxin, 1926: Études du pH interne des larves de l'Oursin *Paracentrotus lividus* pendant la régression du squelette déterminée par l'acidification du milieu extérieur. C. r. Soc. Biol. **94**, 496—498.
— et R. Wurmser, 1926: Sur le potentiel de réduction du noyau et les oxydations cellulaires. C. r. Soc. Biol. **94**, 989—990.
— — 1928: On intracellular oxidation-reduction potential. Proc. roy. Soc., Lond. **102 B**, 128—136.
Raven, C. P., 1937: Experimentelle Untersuchungen über die „Bipolare Differenzierung" des Polychaeten- und Molluskeneis. Acta Neerl. Morphol. **1**, 337—357.
Reilly, J., and W. N. Rae, 1953: Physico-Chemical Methods. Volume 2. New York.
Reiss, P., 1924 a: Quelques données sur le pH intérieur apparent du protoplasme et du noyau. Arch. Physique biol. **4**, 33—42.
— 1924 b: Remarques sur le pH intérieur du noyau cellulaire et ses variations expérimentales. C. r. acad. sci. **179**, 641—643.
— 1925: L'évolution du pH intérieur de l'Oursin pendant la fécondation et la division. C. r. Acad. Paris **181**, 936—938.
— 1926 a: Le pH Intérieur Cellulaire. Paris.
— 1926 b: La réduction des indicateurs comme cause d'erreur des mesures colorimétrique du pH. C. r. Soc. Biol. **94**, 289.
— 1928 a: Étude du vert de bromcrésol comme indicateur de pH intérieur cellulaire. Bull. Inst. Oceanogr. Monaco, No. 526, 1—14.
— 1928 b: Étude du vert de bromcrésol comme indicateur de pH intérieur cellulaire. Arch. Physique biol. **7**, 25—38.
— und M. Gersch, 1936: Die Zelldifferenzierung und Zellspezialisierung während der Embryonalentwicklung von *Aplysia limacina* L., zugleich ein Beitrag zu Problemen der vitalen Färbung. Publ. Staz. Napoli **15**, 223—273.
— et E. Vellinger, 1926: Recherches potentiométriques sur le pH intérieur du muscle. C. r. Soc. Biol. **94**, 1368—1371.
Ricci, J. E., 1952: Hydrogen Ion Concentration. New Concepts in a Systematic Treatment. Princeton.
Rous, P., 1924: The relative reaction of living mammalian tissues. Science **60**, 363—364.
— 1925 a: General features of vital staining with litmus. J. exper. Med. (Am.) **41**, 379—398.
— 1925 b: On the mobilization of acid material within cells and the reaction as influenced by the cell state. J. exper. Med. (Am.) **41**, 399—411.
— 1925 c: Indicated differences in the reaction of the blood and tissues on vital staining with phthaleins. J. exper. Med. (Am.) **41**, 451—470.
— 1925 d: Indicated differences in the reaction of the organs on vital staining with the phthaleins. J. exper. Med. (Am.) **41**, 739—759.
— 1926: Factors determining the reaction of skin grafts; a study by the indicator method of conditions within an ischemic tissue. J. exper. Med. (Am.) **44**, 815—834.
— and W. W. Beattie, 1926: The influence in changes in reaction of the blood upon the reaction of the tissues. J. exper. Med. (Am.) **44**, 835—854.

Rumjantzew, A., und B. Kedrowsky, 1926: Untersuchung über Vitalfärbung einiger Protisten. Protoplasma **1**, 189—203.

Runnström, J., 1949: The mechanism of fertilization in the Metazoa. Advances in Enzymol. **9**, 241—347.

Saito, Y., 1936: Investigations on the biological characters of *Leishmania donovani*. J. orient. Med. (Mandsch.) **24**, 43.

Scarth, G. W., 1924: Can the hydrogen ion concentration of living protoplasm be determined? Science **60**, 431—432.

Schade, H., P. Neukirch und A. Halpert, 1921: Über lokale Acidosen des Gewebes und die Methodik ihrer intravitalen Messung, zugleich ein Beitrag zur Lehre der Entzündung. Z. exper. Med. **24**, 11—56.

Schmidtmann, M., 1924: Über eine Methode zur Bestimmung der Wasserstoffzahl im Gewebe und in einzelnen Zellen. Biochem. Z. **150**, 253—255.

— 1925 a: Über die intrazelluläre Wasserstoffionenkonzentration. Klin. Wschr. **4**, 759.

— 1925 b: Über die intrazelluläre Wasserstoffionenkonzentration unter physiologischen und einigen pathologischen Bedingungen. Z. exper. Med. **45**, 714—742.

— 1927 a: Über intrazelluläre Wasserstoffionenkonzentration und ihre praktische Bedeutung. Z. ärztl. Fortbild. **24**, 50—54.

— 1927 b: Über intrazelluläre Wasserstoffionenkonzentration zur Methodik. Z. exper. Med. **57**, 123—144.

— und K. Matthes, 1927: Untersuchungen über die Reaktion und Permeabilität von Zellen des entzündeten und Geschwulstgewebes. Z. exper. Med. **57**, 127—144.

— und A. Seki-Nureddin, 1927: Über intrazelluläre Wasserstoffionenkonzentration: Beeinflussung der Wasserstoffionenkonzentration von Organzellen im Durchspülungsversuch. Z. exper. Med. **58**, 340—361.

Shapiro, N. N., 1927: The cycle of hydrogen ion concentration in the food vacuoles of *Paramecium, Vorticella* and *Stylonychia*. Trans. amer. Microsc. Soc. **46**, 45—53.

Small, J., 1952: The new outlook on pH. Protoplasma **151**, 273—276.

— 1955: The pH of plant cells. Protoplasmatologia II. B. 2. c. Vienna.

Sörensen, S. P. L., 1909 a: Über die Messung und die Bedeutung der Wasserstoffionenkonzentration bei enzymatischen Prozessen. Biochem. Z. **21**, 131—304.

— 1909 b: Étude enzymatiques. II. Sur la mesure et l'importance de la concentration des ions hydrogène dans les réactions enzymatiques. C. r. Lab. Carlsberg **8**, 1—165.

Spek, J., 1933: Die bipolare Differenzierung des Protoplasmas des Telosteer-Eies und ihre Entstehung. Protoplasma **18**, 497—544.

— 1934 a: Über die bipolare Differenzierung der Eizellen von *Nereis limbata* und *Chaetopterus pergamentaceus*. Protoplasma **21**, 394—405.

— 1934 b: Die Reaktion der Protoplasma-Komponenten des *Asterias*-Eies. Protoplasma **21**, 561—576.

— 1934 c: Die bipolare Differenzierung des Cephalopoden- und des Prosobranchiereies. Vitalfärbungsversuche mit Indikatoren an den Eiern von *Loligo vulgaris* und *Columbella avara*. Arch. Entw.mechan. **131**, 362—372.

— 1937: Das pH in der lebenden Zelle. Erg. Enzymforsch. **6**, 1—20.

— 1940: Metachromasie und Vitalfärbung mit pH-Indikatoren. Protoplasma **34**, 533—584.

— und R. Chambers, 1933: Das Problem der Reaktion des Protoplasmas. (Neue experimentelle Studien, ausgeführt an Amöben.) Protoplasma **20**, 376—406.

Stella, G., 1929: The combination of carbon dioxide with muscle: its heat of neutralization and its dissociation curve. J. Physiol. (Brit.) **68**, 49—66.

Strelnikov, S. D., 1929: L'absorption des colorants basiques par *Paramecium caudatum*. C. r. Soc. Biol. **100**, 1004—1006.

Taylor, H. S., 1931: Treatise on Physical Chemistry. 2nd Edition. New York.

Tchakhotine, S., 1929: Il pH intracellulare dell'uovo de ricci di mare determinate con metodo della microraggiopunctura. Boll. Soc. ital. Biol. sper. **4**, 480—483.

Tiffeneau, M., et D. Broun, 1935: Influence opposée des ions H et OH sur les actions pharmacodynamiques, concernant les appareils autonomes, action sur le tonus utérin. C. r. Soc. Biol. **119**, 1380—1382.

Tomicek, O., 1951: Chemical Indicators. Translated by A. R. Weir. London.

Trifonova, A. N., 1934: Parthenogenese der Fische. Acta Zool. **15**, 183—213.
— 1937: La physiologie de la différentiation et de la croissance. 1. L'équilibre Pasteur-Meyerhoff dans le développement des Poissons. Acta Zool. **18**, 375—445.
— M. F. Vernidoube et N. D. Phillippov, 1939: La physiologie de la différentiation et de la croissance. II. Les périodes critiques dans le développement des Salmonidés et leur base physiologique. Acta Zool. **20**, 239—267.
Vellinger, E., 1926: Recherches potentiométriques sur le pH intérieur de l'œuf d'Oursin. C. r. Soc. Biol. **94**, 1371—1373.
— 1927: Recherches potentiométriques sur le pH intérieur et sur le potential d'oxydation-réduction de l'œuf d'Oursin. Bull. Inst. océanograph. Monaco **506**, 1—12.
— 1928: Notes sur l'électrode d'antimoine. III. L'erreur saline et l'erreur protéique. Arch. Physique biol. **7**, 39—51.
Vlès, F., 1924: Recherches sur le pH intérieur cellulaire. Arch. Physique biol. **4**, 1—20.
— 1925 a: Considérations théoriques sur le point isoélectrique des ampholytes: Leur application à la formation de complexes. Arch. Physique biol. **4**, 228—284.
— 1925 b: Notes sur la mesure spectrophotométrique du pH. Arch. Physique biol. **4**, 285—321.
— 1926: Microcolorimétrie pour les mesures microscopiques de pH ou de rH. C. r. Soc. Biol. **94**, 879—881.
— G. Achard et D. Prikelmaier, 1923: Sur quelques propriétés physico-chimiques des constituants de l'œuf d'Oursin. C. r. acad. sci. **176**, 1179—1181.
— et A. de Coulon, 1924: Recherches sur les propriétés physico-chimiques des tissus en relation avec l'état normal ou pathologique de l'organisme. C. r. acad. sci. **179**, 82—85.
— — 1925: Recherches sur les propriétés physico-chimiques des tissus en relation avec l'état normal ou pathologique de l'organisme. Arch. Physique biol. **4**, 43—85.
— — 1928: Nouvelle expériences sur les courbes de réceptivité de la Souris pour les greffes de tumeurs. Arch. Physique biol. **7**, 1—24.
— et M. Gex, 1928: Recherches sur le spectre ultraviolet de l'œuf d'Oursin (*Paracentrotus lividus* Lk.) et de ses constituants. Bull. inst. océanograph. Monaco **518**, 1—31.
— P. Reiss et E. Vellinger, 1924 a: Mesures potentiométriques sur le pH de substances des œufs d'Oursin. C. r. acad. sci. **179**, 349—351.
— — — 1924 b: Recherches potentiométriques sur le pH intérieur de l'œuf d'Oursin. Bull. inst. océanograph. Monaco **450**, 1—14.
— — — 1924 c: Recherches potentiométriques sur le pH intérieur de l'œuf d'Oursin. Arch. Physique biol. **4**, 21—42.
— et E. Vellinger, 1928: Recherches sur le pigment de l'œuf d'*Arbacia* envisagé comme indicateur de pH intracellulaire. Bull. inst. océanograph. Monaco **513**, 1—16.
Vonwiller, P., 1921: Intravitale Färbung von Protozoa. Abderhaldens Handbuch der biol. Arbeitsmethoden **5**, 87—96.
Walther, O. A., et J. Ulrich, 1926: Une microméthode colorimétrique de mesure du pH. Bull. Soc. chim. biol. **8**, 994 et 1106—1111.
Wiercinski, F. J.. 1941: An experimental study of intracellular pH in the *Arbacia* egg. Biol. Bull. (Am.) **81**, 303.
— 1944: An experimental study of protoplasmic pH determination. 1. *Amoebae* and *Arbacia punctulata.* Biol. Bull. (Am.) **86**, 98—112.
— 1952: The relation of pH to the shortening of muscle protoplasm by cations. Fed. Proc. no. **1**, 11, 172.
Willmer, E. N., 1944: Colour of small objects. Nature **153**, 774—775.
Yamamoto, T., 1936: Studies on the rythmical movements of the early embryo of *Oryzias latipes.* 7. Anaerobic movements and oxidation-reduction potential of the egg limiting the rythmical movements. Fac. Sci. Imp. Univ. Tokyo **4**, 233—248.
Yosezato, M., 1933: Hydrogen-ion concentration of *Entamoeba histolytica.* J. orient. Med. (Mandsch.) **19**, 66.

GPSR Compliance

The European Union's (EU) General Product Safety Regulation (GPSR) is a set of rules that requires consumer products to be safe and our obligations to ensure this.

If you have any concerns about our products, you can contact us on ProductSafety@springernature.com

In case Publisher is established outside the EU, the EU authorized representative is:

Springer Nature Customer Service Center GmbH
Europaplatz 3
69115 Heidelberg, Germany

Batch number: 10578397

Printed by Printforce, the Netherlands